Lecture Notes in Mathematics

Edited by A. Dold and B. Eckmann

Subseries: Department of Mathematics, University of Maryland
Adviser: J.K. Goldhaber

227

Hugh L. Montgomery

Topics in Multiplicative Number Theory

Springer-Verlag
Berlin Heidelberg New York Tokyo

Author

Hugh L. Montgomery
Mathematics Department, University of Michigan
Ann Arbor, MI 48109, USA

1st Edition 1971
2nd Printing 1986

Mathematics Subject Classification (1980): 10Hxx, 10J15

ISBN 978-3-540-05641-6

Printing and binding: Beltz Offsetdruck, Hemsbach/Bergstr.
2146/3140-543210

P R E F A C E

These Notes are designed to present a survey of
the present state of knowledge concerning those subjects
touched upon in the last fifty pages of Davenport's
Multiplicative Number Theory. Davenport's book is an
admirable introduction to the subject at hand; most
results prerequisite to these Notes are found in §§ 1-22
of his book.

The more general results in this area are found in
the first chapters. Applications are then made to the
zeta-function and L-functions, and finally these results
are used to derive theorems concerning the distribution
of prime numbers.

Due to continuing research in this field, these Notes
are already a little out of date. In particular, Jutila
has sharpened my estimate (12.14) in Theorem 12.2 and I
am now in a position to say more concerning my Conjecture
9.2.

I am indebted to many people for their suggestions
and remarks. In addition, E. Bombieri, P. X. Gallagher,
G. Halász, M. H. Huxley, A. Selberg, P. Turán and R.
Vaughan have made available to me valuable unpublished
material. My presentation has benefited from the specific
suggestions and criticisms of Dr. Baker, Professor Bateman,
and Professor Davenport. While researching and writing
I have been grateful to receive monetary support from the
Marshall Aid Commemoration Commission (London), the

National Science Foundation (Washington), Trinity College
(Cambridge), the Institute for Advanced Study (Princeton),
and the Air Force Office of Scientific Research (Washington).
My thanks also go to Mrs. J. M. Jones for her diligent
typing of my manuscript.

H. L. M.

LOGICAL STRUCTURE

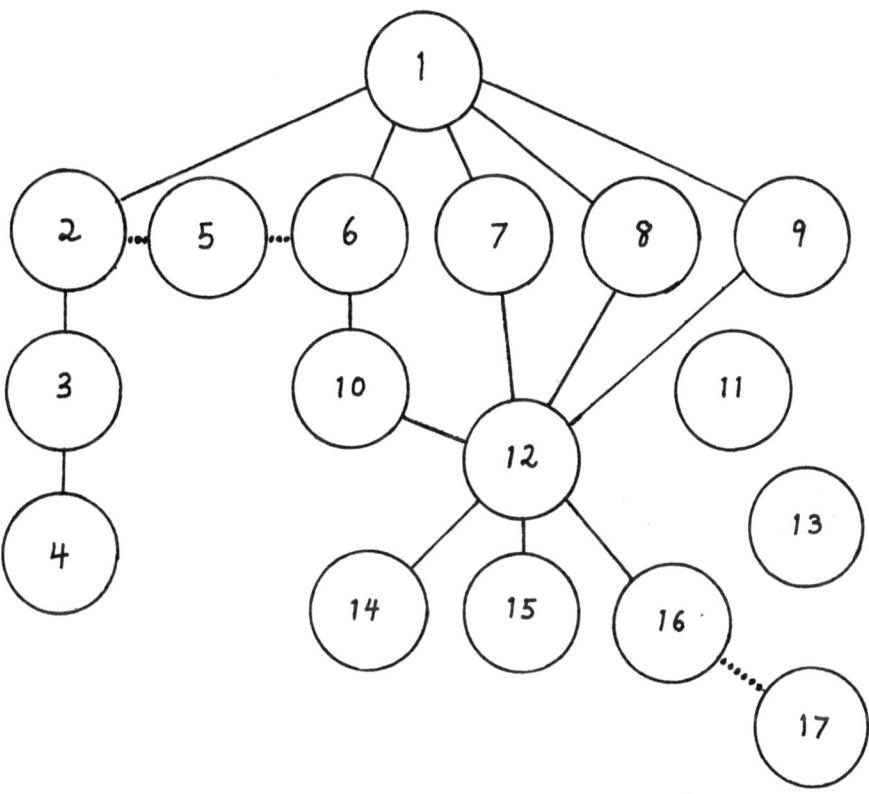

Solid lines indicate that some result in the later chapter depends on some result in the earlier chapter. Chapters 2, 5, 6 share common ideas, but the results are not interdependent. Lemma 17.3, which is quoted from the literature, could be proved in such a manner that it would depend on the results of Chapter 16.

NOTATION

We employ the familiar notations $O(\quad)$, $o(\quad)$, $\Omega(\quad)$, $\Omega_{\pm}(\quad)$, $\sim$. In addition we use further conventions introduced by I.M. Vinogradov: If f and g are two functions and $g \geq 0$, then we write $f = O(g)$ or $f \ll g$ if there is an absolute constant A such that $|f| \leq Ag$. The expression $g \gg f$ means the same as $f \ll g$, but we write $g \gg f$ only when both f and g are non-negative. If an estimate is not uniform with respect to certain parameters then these parameters are explicitly mentioned or are included in the expression as subscripts, e.g. $f = O_{\varepsilon}(g)$.

We write $[x]$ for the greatest integer not exceeding x, and we write $\|x\|$ for the distance from x to the nearest integer, so

$$\|x\| = \min(x - [x], 1 + [x] - x) = \min_{n \in \mathbb{Z}} |x - n|.$$

We let $e(x)$ denote the function

$$e(x) = e^{2\pi i x}.$$

The letters n and k are generally positive or non-negative integers, p is a prime, and q is a general modulus. The function $\chi(n)$ is a character modulo q, and χ^{*} is the primitive character which induces χ. The arithmetic functions of Euler, Möbius, and von Mangoldt we write as usual; we use $d(n)$ for the divisor function. We let $\pi(x)$

be the usual counting function of the primes, and $\psi(x)$ the counting function of Chebychev,

$$\psi(x) = \sum_{n \leq x} \Lambda(n).$$

As elsewhere, and in particular as in Davenport's book, we let

$$\psi(x; q, a) = \sum_{\substack{n \leq x \\ n \equiv a \pmod{q}}} \Lambda(n)$$

and

$$\psi(x, \chi) = \sum_{n \leq x} \Lambda(n) \chi(n).$$

We let $s = \sigma + it$ be a complex variable with real part σ and imaginary part t. Similarly $\rho = \beta + i\gamma$ is a non-trivial zero of $\zeta(s)$ or of a certain L-function. We write $\sum_{\rho}$ to denote a sum of over all non-trivial zeros of a particular L-function. We let $\sum_{\chi}$ denote a sum over all characters modulo q, and let $\sum_{\chi}^{*}$ denote a sum over all primitive characters modulo q. We write $\sum_{a=1}^{q}{}^{*}$ as a shorthand for $\sum_{\substack{a=1 \\ (a,q)=1}}^{q}$.

CONTENTS

Contents

Three basic principles

In Chapters 2, 5, 6, 7, 8 and 9 we discuss a variety of
results whose derivations depend on only three basic ideas.
Here we discuss these ideas in detail, and formulate them in
several lemmas for easy use.

Our first idea has been implicit in the literature for
some time (see Littlewood's Miscellany, p. 36, and §7.65 of
Titchmarsh's Theory of Functions). In its simplest form it
is

LEMMA 1.1 (Sobolev - Gallagher). Let $a < b$ be real
numbers, and f a continuous complex valued function on $[a, b]$,
with continuous first derivative in (a, b). Then

$$|f(\tfrac{a+b}{2})| \leq (b-a)^{-1} \int_a^b |f(x)|\,dx + \tfrac{1}{2}\int_a^b |f'(x)|\,dx, \qquad (1.1)$$

and

$$|f(u)| \leq (b-a)^{-1} \int_a^b |f(x)|\,dx + \int_a^b |f'(x)|\,dx \qquad (1.2)$$

for any u in $[a, b]$.

The inequality (1.1) is due to Gallagher [68], who also
recognized its pertinence to analytic number theory. The
second inequality is a special case of an inequality of Sobolev
which in the general case deals with a function of n real
variables which possesses continuous partial derivatives of

order n. We note that (1.2) contains (1.1), except for a slight weakening of the constant. By repeated application of (1.1) Gallagher obtained a result, which (slightly adapted) is

LEMMA 1.2 (Gallagher). Let T_0 and $T \geq \delta > 0$ be real numbers, and let f be a continuous complex valued function on the interval $[T_0, T_0 + T]$, with continuous derivative in $(T_0, T_0 + T)$. Let $\mathcal{T}$ be a set of real numbers in the interval $[T_0 + \frac{\delta}{2}, T_0 + T - \frac{\delta}{2}]$, and suppose that

$$|t - t'| \geq \delta \tag{1.3}$$

for distinct t, t' in $\mathcal{T}$. Then

$$\sum_{t \in \mathcal{T}} |f(t)| \leq \delta^{-1} \int_{T_0}^{T_0 + T} |f(x)| \, dx + \frac{1}{2} \int_{T_0}^{T_0 + T} |f'(x)| \, dx. \tag{1.4}$$

We note that the hypothesis (1.3) may be omitted, but the result we obtain is more complicated. If $\eta > 0$ and $\mathcal{T}$ is finite then we let

$$N_\eta(x) = \sum_{\substack{t \in \mathcal{T} \\ |t - x| < \eta}} 1. \tag{1.5}$$

We have

LEMMA 1.3. Let $T_0, T \geq \delta > 0$ be real numbers, and let $\mathcal{T}$ be a finite set in the interval $[T_0 + \frac{\delta}{2}, T_0 + T - \frac{\delta}{2}]$. With $N_\eta(x)$ defined by (1.5) and f as in Lemma 1.2 we have

$$\sum_{t \in \mathcal{T}} N_\delta(t)^{-1} |f(t)| \leq \delta^{-1} \int_{T_0}^{T_0 + T} |f(x)| \, dx + \frac{1}{2} \int_{T_0}^{T_0 + T} |f'(x)| \, dx. \tag{1.6}$$

If (1.3) holds then $N_\delta(t) = 1$ for $t \in \mathcal{T}$, so the above contains Lemma 1.2, without any loss in constants.

In most of our applications f is the square of a function, $f(t) = S(t)^2$, so $f'(t) = 2 S(t) S'(t)$. From Lemma 1.3 and the Cauchy - Schwarz inequality we have immediately

LEMMA 1.4 (Gallagher). <u>Let</u> T_0, T, δ, $\mathcal{T}$, <u>and</u> $N_\delta(t)$ <u>be</u> <u>as in Lemma 1.3, and suppose that</u> S <u>is a continuous complex</u> <u>valued function in</u> $[T_0, T_0 + T]$ <u>with continuous derivative in</u> $(T_0, T_0 + T)$. Then

$$\sum_{t \in \mathcal{T}} N_\delta(t)^{-1} |S(t)|^2 \leq \delta^{-1} \int_{T_0}^{T_0+T} |S(t)|^2 dt + \left(\int_{T_0}^{T_0+T} |S(t)|^2 dt \right)^{\frac{1}{2}} \left(\int_{T_0}^{T_0+T} |S'(t)|^2 dt \right)^{\frac{1}{2}} \cdot \quad (1.7)$$

We shall find the above inequality to be very useful, because it allows one to replace an integral by a sum over well-spaced points.

We now prove Lemma 1.1. By integration by parts one can easily verify that

$$f(u) = (b-a)^{-1} \int_a^b f(x) dx + \int_u^b \left(\frac{x-b}{b-a} \right) f'(x) dx + \int_a^u \left(\frac{x-a}{b-a} \right) f'(x) dx, \quad (1.8)$$

for $u \in [a, b]$. This makes (1.2) trivial. We see that (1.1) also follows, on taking $u = \frac{1}{2}(a+b)$.

We now prove Lemma 1.3. From (1.1) we see that

$$|f(t)| \leq \delta^{-1} \int_{t-\frac{\delta}{2}}^{t+\frac{\delta}{2}} |f(x)| dx + \frac{1}{2} \int_{t-\frac{\delta}{2}}^{t+\frac{\delta}{2}} |f'(x)| dx,$$

so

$$\sum_{t \in \mathcal{T}} N_\delta(t)^{-1} |f(t)| \leq \delta^{-1} \int_{T_0}^{T_0+T} |f(x)| w(x) dx + \frac{1}{2} \int_{T_0}^{T_0+T} |f'(x)| w(x) dx,$$

where

$$w^-(t) = \sum_{\substack{t \in \mathcal{T} \\ |t-x| < \frac{\delta}{2}}} N_\delta(t)^{-1}.$$

It suffices to show that $w^-(t) \leq 1$. Now $N_\delta(t)$ counts precisely those $t' \in \mathcal{T}$ for which $|t'-t| < \delta$, so if $|t-x| < \frac{\delta}{2}$ then $N_\delta(t)$ counts at least those $t' \in \mathcal{T}$ for which $|t'-x| < \frac{\delta}{2}$, which implies that $N_\delta(t) \geq N_{\frac{\delta}{2}}(x)$ for $|t-x| < \frac{\delta}{2}$. Hence

$$w^-(x) \leq N_{\frac{\delta}{2}}(x)^{-1} \sum_{\substack{t \in \mathcal{T} \\ |t-x| < \frac{\delta}{2}}} 1$$

$$= 1.$$

This completes the proof of Lemma 1.3. Of course if (1.3) holds then it is obvious that $w^-(t) \leq 1$, as the intervals of integration are non-overlapping in this case.

We now discuss our second main tool, which is a generalization of Bessel's inequality for vectors in an inner product space. To fix our ideas we recall that if $\varphi_1, \varphi_2, \ldots, \varphi_R$ are orthonormal elements of an inner product space over the complex numbers then for any ξ we have

$$\sum_{r=1}^{R} |(\xi, \varphi_r)|^2 \leq \|\xi\|^2. \tag{1.9}$$

This is Bessel's inequality, which is easily proved (see §63 of Halmos' book). If $R = 1$ then we have

$$|(\xi, \zeta)| \leq \|\xi\| \|\zeta\| \tag{1.10}$$

for any ξ and ζ; this is Schwarz's inequality. We desire an inequality similar to (1.9), but one which applies even when the φ_r are not necessarily orthonormal. Boas [11] has given a number of results of this sort, and Bellman [9] later rediscovered one of Boas' most interesting results, which states that

$$\sum_{r=1}^{R} |(\xi, \varphi_r)|^2 \leq \|\xi\|^2 \left(\max_{1 \leq r \leq R} \|\varphi_r\| + \left(\sum_{r \neq s} |(\varphi_r, \varphi_s)|^2 \right)^{\frac{1}{2}} \right) \quad (1.11)$$

for any $\xi, \varphi_1, \varphi_2, \ldots, \varphi_R$. We note that if the φ_r happen to be orthonormal then the second factor on the right hand side is 1; hence (1.11) includes Bessel's inequality (1.9). Recently Bombieri [19] found the following variation of (1.11).

LEMMA 1.5 (Bombieri). If $\xi, \varphi_1, \varphi_2, \ldots, \varphi_R$ are elements of an inner product space over the field of complex numbers, then

$$\sum_{r=1}^{R} |(\xi, \varphi_r)|^2 \leq \|\xi\|^2 \max_{1 \leq r \leq R} \sum_{s=1}^{R} |(\varphi_r, \varphi_s)|. \quad (1.12)$$

We see that if the φ_r are orthonormal then the above also reduces to (1.9).

In fact Bombieri's lemma is equivalent to

LEMMA 1.6. If $\xi, \varphi_1, \varphi_2, \ldots, \varphi_R$ are as above, then

$$\left| \sum_{r=1}^{R} c_r (\xi, \varphi_r) \right| \leq \left(\sum_{r=1}^{R} |c_r|^2 \right)^{\frac{1}{2}} \|\xi\| \left(\max_{1 \leq r \leq R} \sum_{s=1}^{R} |(\varphi_r, \varphi_s)| \right)^{\frac{1}{2}}, \quad (1.13)$$

where the c_r are arbitrary real or complex numbers.

To derive (1.13) from (1.12) we have only to use Cauchy's inequality,

$$\left| \sum_{r=1}^{R} c_r (\xi, \varphi_r) \right| \leq \left(\sum_{r=1}^{R} |c_r|^2 \right)^{\frac{1}{2}} \left(\sum_{r=1}^{R} |(\xi, \varphi_r)|^2 \right)^{\frac{1}{2}},$$

and then use (1.12) to majorize to the second factor on the
right hand side.   To derive (1.12) from (1.13) it suffices
to take $c_r = \overline{(\xi, \varphi_r)}$, divide out the common factor
$\left(\sum_r |(\xi, \varphi_r)|^2 \right)^{\frac{1}{2}}$, and square both sides.

The following gives an abstract formulation of an
inequality of Halász [73], [74] , [75] (see also [142]).
The inequality of Heilbronn [85], when abstracted, is weaker
than what follows.

LEMMA 1.7.   Let ξ, $\varphi_1, \varphi_2, \ldots, \varphi_R$ be as above.   Then
$$\sum_{r=1}^{R} |(\xi, \varphi_r)| \leq \|\xi\| \left(\sum_{r, s} |(\varphi_r, \varphi_s)| \right)^{\frac{1}{2}} . \tag{1.14}$$

We note that this is slightly stronger than (1.13) with
$c_r = exp(-i \arg(\xi, \varphi_r))$.

Rényi [156], [157], [158], [159], [160], [161] (see also
Chapter 10 of Halberstam and Roth) was the first to realize
that inequalities like (1.11), (1.12), or (1.15) could be used
in proving the large sieve.   On the other hand Halász and
Turán [175] have used Halász's method to obtain new results in
the theory of the zeta function, while Bombieri has observed
that one can derive a modified form of Halász's inequality
from (1.12).   In Chapters 8 and 9 we shall discuss the
implications of (1.12) concerning arbitrary Dirichlet
polynomials.

Independently of Bombieri, A. Selberg has found a slightly

stronger result, which we state as

LEMMA 1.8 (Selberg). <u>If</u> ξ, φ_1, φ_2, ..., φ_R <u>are</u> <u>as</u> <u>above</u>

<u>then</u>

$$\sum_{r=1}^{R} |(\xi, \varphi_r)|^2 \left(\sum_{s=1}^{R} |(\varphi_r, \varphi_s)|\right)^{-1} \leq \|\xi\|^2. \tag{1.15}$$

We now prove Lemma 1.6, and in consequence, Lemma 1.5.
We have

$$\sum_{r=1}^{R} c_r(\xi, \varphi_r) = \left(\xi, \sum_{r=1}^{R} \bar{c}_r \varphi_r\right),$$

so by Schwarz's inequality (1.10)

$$\left|\sum_{r=1}^{R} c_r(\xi, \varphi_r)\right| \leq \|\xi\| \|\sum \bar{c}_r \varphi_r\|. \tag{1.16}$$

The square of the second factor is

$$\sum_{r} \sum_{s} \bar{c}_r c_s (\varphi_r, \varphi_s), \tag{1.17}$$

which is a Hermitian form in the variables c_r. It remains
to bound this form. We have $|\bar{c}_r c_s| \leq \frac{1}{2}(|c_r|^2 + |c_s|^2)$,
so

$$\sum_{r} \sum_{s} \bar{c}_r c_s (\varphi_r, \varphi_s) \leq \sum_{r} |c_r|^2 \sum_{s=1}^{R} |(\varphi_r, \varphi_s)| \tag{1.18}$$

$$\leq \left(\sum_{r} |c_r|^2\right) \max_{1 \leq r \leq R} \sum_{s=1}^{R} |(\varphi_r, \varphi_s)|.$$

This, with (1.16), gives (1.13), so the proof is complete.
We have followed Bombieri's proof of Lemma 1.5, except that
Bombieri's treatment of the Hermitian form (1.17) involved a

discussion of its eigenvalues.   The simplifications above were suggested to me by the proofs of Theorems 273, 274, 275 in the book of Hardy, Littlewood, and Pólya, and by a remark of Wirsing.

To prove Lemma 1.7 we follow the above proof with $c_r = exp\left(-i \, arg\left(\xi, \varphi_r\right)\right)$, but we now give a simpler bound for (1.17), namely

$$\sum_r \sum_s \overline{c_r} \, c_s \, (\varphi_r, \varphi_s) \;\leq\; \sum_{r,s} |(\varphi_r, \varphi_s)| .$$

This, with (1.16), gives (1.14).

We now give Selberg's proof of Lemma 1.8.   We have

$$\| \xi - \sum_r c_r \varphi_r \|^2 \geq 0$$

for any complex numbers c_r, that is to say

$$\| \xi \|^2 - 2 Re \sum_r \overline{c_r} (\xi, \varphi_r) + \sum_{r,s} c_r \overline{c_s} (\varphi_r, \varphi_s) \;\geq\; 0 .$$

From (1.18) we have

$$2 Re \sum_r \overline{c_r} (\xi, \varphi_r) \;\leq\; \| \xi \|^2 + \sum_r |c_r|^2 \sum_s |(\varphi_r, \varphi_s)| .$$

If we now take $c_r = (\xi, \varphi_r)\left(\sum_s |(\varphi_r, \varphi_s)|\right)^{-1}$ the result follows.

Our third main tool is another invention of Gallagher [71]. We let

$$S(t) = \sum_{\mu \in \mathcal{M}} c(\mu) \, e(\mu t), \qquad\qquad (1.19)$$

where $\mathcal{M}$ is a certain countable set of real numbers and the $c(\mu)$ are arbitrary real or complex numbers, subject to the condition that

$$\sum_{\mu} | c(\mu) | < \infty . \tag{1.20}$$

We wish to have a bound for the mean square of S, in terms of the coefficients $c(\mu)$. Gallagher's result is

LEMMA 1.9 (Gallagher). For any $\varepsilon > 0$, if δ and T are positive real numbers for which $\delta T \leq 1 - \varepsilon$, and if (1.19) and (1.20) hold, then

$$\int_{-T}^{T} | S(t) |^2 dt \ll_{\varepsilon} \int_{-\infty}^{+\infty} | C_{\delta}(x) |^2 dx, \tag{1.21}$$

where

$$C_{\delta}(x) = \delta^{-1} \sum_{|\mu - x| < \frac{\delta}{2}} c(\mu). \tag{1.22}$$

We use this to give a result for Dirichlet series, as follows.

LEMMA 1.10 (Gallagher). If

$$S(s) = \sum_{n=1}^{\infty} a_n n^{-s} \tag{1.23}$$

is absolutely convergent for $\operatorname{Re} s \geq 0$ then

$$\int_{-T}^{T} | S(t) |^2 dt \ll T^2 \int_{0}^{\infty} | \sum_{y}^{\tau y} a_n |^2 \frac{dy}{y}, \tag{1.24}$$

where $\tau = exp(T^{-1})$ and $T > 0$.

In Chapter 5 we use the idea of the proof of Lemma 1.9 to obtain a bound like (1.24), but with the best constant.

We now give Gallagher's proof of Lemma 1.9. Let $F_{\delta}(x)$

be δ^{-1} or 0, according as $|x| < \frac{\delta}{2}$ or not. Then

$$C_\delta(x) = \sum_\mu c(\mu) F_\delta(x-\mu).$$

The Fourier transform of C_δ is

$$\hat{C}_\delta(t) = \int_{-\infty}^{+\infty} C_\delta(x) e(xt) dx$$

$$= \sum_\mu c(\mu) \int_{-\infty}^{+\infty} F_\delta(x-\mu) e(xt) dx$$

$$= \sum_\mu c(\mu) e(\mu t) \int_{-\infty}^{+\infty} F_\delta(x-\mu) e((x-\mu)t) dx$$

$$= S(t) \hat{F}_\delta(t).$$

Consequently, by Plancherel's theorem,

$$\int_{-\infty}^{+\infty} |S(t) \hat{F}_\delta(t)|^2 dt = \int_{-\infty}^{+\infty} |C_\delta(x)|^2 dx. \qquad (1.25)$$

But $\hat{F}_\delta(t) = \frac{\sin \pi \delta t}{\pi \delta t}$, so if $\delta T \leq 1 - \varepsilon$ then $|\hat{F}_\delta(t)| \gg_\varepsilon 1$ for $|t| \leq T$, and the result follows.

To obtain Lemma 1.10 from Lemma 1.9 we take $\mu = \mu(n) = -(2\pi)^{-1} \log n$, $x = -(2\pi)^{-1} \log y$, and $\delta = (2\pi T)^{-1}$. Then from (1.21) we have

$$\int_{-T}^{T} |S(t)|^2 dt \ll T^2 \int_0^\infty \left| \sum_{y\tau^{-\frac{1}{2}}}^{y\tau^{\frac{1}{2}}} a_n \right|^2 \frac{dy}{y}$$

$$= T^2 \int_0^\infty \left| \sum_y^{\tau y} a_n \right|^2 \frac{dy}{y}.$$

C H A P T E R   2

The large sieve

The large sieve has its origins in the works of
Linnik [122] and Rényi [155], [156], [157], [158], [159],
[160], [161] (see also Chapter IV §10 of Halberstam and Roth).
On this foundation the work of Roth [177], [179] was a
considerable advance, and this was soon followed by an
important paper of Bombieri [15].   Davenport and Halberstam [47]
were the first to discuss the large sieve in its most basic
form, namely as an upper bound for

$$\sum_{x \in \mathcal{X}} |S(x)|^2, \qquad\qquad (2.1)$$

where $\mathcal{X}$ is a set of real numbers (well-spaced modulo 1) and $S(x)$
is an arbitrary exponential polynomial,

$$S(x) = \sum_{n=M+1}^{M+N} a_n e(nx). \qquad\qquad (2.2)$$

Bounds for (2.1) have since been given by Bombieri and Davenport
[20], [21], Gallagher [68], Ming-Chit Liu [140], and
Bombieri [18], while Rieger [165], [166], Huxley [91], [93],
Wilson [227], Schaal [193], Samandarov [182], Hlawka [86],
and Johnsen   have given a variety of generalizations.
Moreover Davenport (see [142]) and Rényi have produced analogues
for Dirichlet polynomials and ordinary polynomials, respectively.

From Lemma 1.4 we have immediately

THEOREM 2.1.   If $S(x)$ is defined by (2.2), $\mathcal{X}$ is a finite set of real numbers in $[0, 1)$, and if $0 < \delta \leq \frac{1}{2}$, then

$$\sum_{x \in \mathcal{X}} N_\delta(x)^{-1} |S(x)|^2 \leq \left(\delta^{-1} + \pi N \right) \sum_n |a_n|^2, \tag{2.3}$$

where

$$N_\delta(y) = \sum_{\substack{x \in \mathcal{X} \\ \|x - y\| < \delta}} 1. \tag{2.4}$$

Davenport [46, §23, Theorem 1B] previously gave (2.3) with an implied constant, and subject to the hypothesis that $\delta \leq (4N)^{-1}$.   Immediately from the above we have

COROLLARY 2.2 (Gallagher).   If $S(x)$ is defined by (2.2) and $\mathcal{X}$ is a set of real numbers for which

$$\|x - x'\| \geq \delta > 0 \tag{2.5}$$

whenever x and x' are distinct members of $\mathcal{X}$, then

$$\sum_{x \in \mathcal{X}} |S(x)|^2 \leq \left(\delta^{-1} + \pi N \right) \sum_n |a_n|^2. \tag{2.6}$$

If the $x \in \mathcal{X}$ are equally spaced mod 1 then $\delta \sum |S(x)|^2$ is a Riemann sum approximating to

$$\int_0^1 |S(x)|^2 dx = \sum |a_n|^2.$$

Hence the bound (2.6) is asymptotically best possible when $N\delta \to 0$.   The size of the secondary term is not best possible, for Bombieri and Davenport [21] have shown that

$$\sum_{x \in \mathcal{X}} |S(x)|^2 \leq \delta^{-1}\left(1 + 270\,(N\delta)^2\right) \sum |a_n|^2. \qquad (2.7)$$

On the other hand Bombieri and Davenport [21] have also produced an $S(x)$ for which there exist $\mathcal{X}$ with δ and $N\delta$ arbitrarily small for which

$$\sum |S(x)|^2 > \delta^{-1}\left(1 + \tfrac{1}{12}(N\delta)^2\right) \sum |a_n|^2. \qquad (2.8)$$

When $N\delta$ is large it is the constant attached to N which is important in Theorem 2.1. If $\mathcal{X}$ contains only a single point x then from Cauchy's inequality we have

$$|S(x)|^2 \leq \left(\sum_n 1\right)\left(\sum_n |a_n|^2\right) = N \sum |a_n|^2, \qquad (2.9)$$

and equality occurs when the a_n are appropriately chosen. Using Lemma 1.5 we easily prove an inequality which essentially contains (2.9). We have

THEOREM 2.3. Under the hypotheses of Corollary 2.2 we have

$$\sum_{x \in \mathcal{X}} |S(x)|^2 \leq \left(N + \tfrac{2}{\sqrt{3}}\delta^{-1} + 3\right) \sum |a_n|^2. \qquad (2.10)$$

We note that if $\mathcal{X}$ contains only one point then (2.5) is vacuous so δ may be taken arbitrarily large. Nevertheless (2.10) remains true, in view of (2.9). Hence we restrict our attention to sets $\mathcal{X}$ which contain at least two elements, and we have $\delta \leq \tfrac{1}{2}$.

For analytic purposes the bound (2.10) has no advantage over (2.6). However, for the arithmetic applications in the next chapter the constant attached to N is of some interest.

Previously Bombieri and Davenport [21] proved (2.10), but with the factor $(N + 5\delta^{-1})$; their argument was very elaborate. Using Lemma 1.5 Bombieri [19] recently proved (2.10) with the factor $(N + 2\delta^{-1})$. In proving (2.10) we make only slight changes in Bombieri's argument. Bombieri and Davenport [21] showed that (2.10) is essentially best possible; they gave examples of $S(x)$, N, δ, $\mathfrak{X}$ for which $\delta \to 0$, $N \to \infty$, $N\delta \to \infty$ and

$$\sum_{x \in \mathfrak{X}} |S(x)|^2 = (N - 1 + \delta^{-1}) \sum |a_x|^2. \tag{2.11}$$

So far as I know it is conceivable that (2.10) is true with the factor $N + \delta^{-1}$. This would be pleasing as it would then contain both (2.6) and (2.10). The inequalities (2.6), (2.7), and (2.10) represent all that is known in this direction, except that Bombieri and Davenport [20] have given (2.10) with the factor $(N^{\frac{1}{2}} + \delta^{-\frac{1}{2}})^2$. Independently Ming-Chit Liu proved (2.10) with the factor $2 \max (N, \delta^{-1})$. This latter result is asymptotically best possible when $N\delta \sim 1$.

By averaging in a suitable way one can reduce the size of the secondary term in (2.10), if $N\delta$ is large. As an example of this we have

THEOREM 2.4. <u>Let</u>

$$S(x) = \sum_{-k}^{k} \left(1 - \frac{|k|}{K}\right) a_k e(kx). \tag{2.12}$$

If $\mathcal{X}$ and δ are as in Corollary 2.2 then

$$\sum_{x \in \mathcal{X}} |S(x)|^2 \leq K(1 + (K\delta)^{-2}) \sum_{-K}^{K} (1 - \frac{|k|}{K})|a_k|^2. \qquad (2.13)$$

It is not difficult to give examples in which $K \to \infty$, $\delta \to 0$, $K\delta \to \infty$, and

$$\sum_{x \in \mathcal{X}} |S(x)|^2 = K(1 + \frac{1}{4}(K\delta)^{-2}) \sum_{-K}^{K} (1 - \frac{|k|}{K})|a_k|^2,$$

so (2.13) is essentially best possible.

The large sieve was initially designed for elementary purposes; a discussion of these is postponed to Chapter 3. In 1948 Rényi [155] found that the large sieve could be used to bound certain averages of character sums. A number of such results are now available [20], [47], [68]. The most useful of these is due to Gallagher [68]; his result is

THEOREM 2.5 (Gallagher). Let

$$T(x) = \sum_{n=M+1}^{M+N} a_n \chi(n). \qquad (2.14)$$

Then for $Q \geq 1$

$$\sum_{q \leq Q} \frac{1}{\phi(q)} \sum_{\chi}^{*} |T(x)|^2 \leq (Q^2 + \pi N) \sum |a_n|^2. \qquad (2.15)$$

This result is obtained from Corollary 2.2 by taking a special set $\mathcal{X}$. Remarks of Erdős [58] and Erdős and Rényi [59] indicate that even in this special situation the large sieve is substantially best possible. We shall discuss some other averages of character sums in Chapters 6, 7, 8, and 9.

We now derive Theorem 2.1. Our main tool is Lemma 1.4.
We first suppose that $S(x)$ is of special shape

$$S(x) = \sum_{-k}^{k} a_k e(kx);$$ (2.16)

we prove that

$$\sum_{x \in \mathcal{X}} N_\delta(x)^{-1} |S(x)|^2 \leq (\delta^{-1} + 2\pi k) \sum_{-k}^{k} |a_k|^2.$$ (2.17)

To complete the proof it will remain to show that (2.3) follows
from (2.17).

We note that as $S(x)$ has period 1 it is not necessary
that $\mathcal{X}$ lie in $\left[\frac{\delta}{2}, 1 - \frac{\delta}{2}\right]$. Rather, $\mathcal{X}$ can lie throughout $[0,1)$
The inequality (1.7) becomes

$$\sum_{x \in \mathcal{X}} N_\delta(x)^{-1} |S(x)|^2 \leq \delta^{-1} \int_0^1 |S(x)|^2 dx + \left(\int_0^1 |S(x)|^2 dx\right)^{\frac{1}{2}} \left(\int_0^1 |S'(x)|^2 dx\right)^{\frac{1}{2}}$$

$$= \delta^{-1} \sum_{-k}^{k} |a_k|^2 + \left(\sum_{-k}^{k} |a_k|^2\right)^{\frac{1}{2}} \left(\sum_{-k}^{k} |k a_k 2\pi|^2\right)^{\frac{1}{2}}$$

$$\leq (\delta^{-1} + 2\pi k) \sum_{-k}^{k} |a_k|^2,$$

and so we have (2.17).

To derive (2.3) from (2.17) we consider two cases. If
N is odd then we write $N = 2k+1$. Then

$$S(x) = e((M+k+1)x) \sum_{-k}^{k} a_{k+k+M+1} e(kx),$$

so (2.17) applies to the sum on the right and (2.3) follows
in this case.   If N is even then we write $a_{M+N+1} = 0$ so the
sum has formally $N+1$ terms.   Then we proceed as above,
as now $N+1 = 2k+1$, the result again follows.

Corollary 2.2 follows from Theorem 2.1, on noting that
(2.5) implies that $N_\delta(x) = 1$ for all $x \in \mathfrak{X}$.

To derive Theorem 2.3 we use Lemma 1.5.   Here as elsewhere,
we take as our space the linear space of square-summable complex
sequences,

$$\underline{\alpha} = \{a_n\}_{-\infty}^{+\infty}, \qquad \sum_{-\infty}^{+\infty} |a_n|^2 < \infty.$$

Our inner product is the usual one,

$$(\underline{\alpha}, \underline{\beta}) = \sum_{-\infty}^{+\infty} a_n \overline{b}_n,$$

where $\underline{\alpha} = \{a_n\}, \underline{\beta} = \{b_n\}.$

As in the proof of Theorem 2.1 we may assume that (2.16)
holds, and then prove that

$$\sum_{x \in \mathfrak{X}} |s(x)|^2 \le \left(2k + \tfrac{2}{\sqrt{3}}\delta^{-1} + 3\right) \sum_{-k}^{k} |a_k|^2. \tag{2.18}$$

In Lemma 1.5 we take

$$\underline{\xi} = \left\{ a_k b_k^{-\frac{1}{2}} \right\}_{-k}^{k}, \qquad \underline{\varphi}_r = \left\{ b_k^{\frac{1}{2}} e(-kx_r) \right\}_{-\infty}^{+\infty},$$

where we index $\mathfrak{X}$ with $r = 1, 2, \ldots, R$.   From (1.12) we have

$$\sum_{x \in \mathcal{X}} |s(x)|^2 \leq \left(\sum_{-K}^{K} |a_k|^2 b_k^{-1} \right) \max_{x \in \mathcal{X}} \sum_{x' \in \mathcal{X}} |B(x'-x)|, \qquad (2.19)$$

where

$$B(x) = \sum_{-\infty}^{+\infty} b_k e(kx), \qquad \sum_{-\infty}^{\infty} b_k < \infty, \qquad (2.20)$$

the b_k are non-negative, and strictly positive for $-K \leq k \leq K$.
The best choice of $B(x)$ remains in doubt; we take

$$b_k = \begin{cases} 1 & \text{for } |k| \leq K, \\ 1 - \frac{|k|-K}{A} & \text{for } K \leq |k| \leq K+A, \\ 0 & \text{for } |k| > K+A, \end{cases} \qquad (2.21)$$

where A is a positive integer to be chosen later. This choice
of the b_k makes $B(x)$ the difference of two Fejér kernels,

$$B(x) = \frac{(\sin(K+A)\pi x)^2 - (\sin K\pi x)^2}{A(\sin \pi x)^2}. \qquad (2.22)$$

From either (2.21) or (2.22) we see that $B(0) = 2N+A$, while
from (2.22) we have $|B(x)| \leq A^{-1}(\sin \pi x)^{-2}$ for non-integral x.
In Appendix I we show that $(\sin \pi x)^{-2} \leq \pi^{-2}\|x\|^{-2} + 1$, so

$$|B(x)| \leq A^{-1}(\pi^{-2}\|x\|^{-2} + 1).$$

From these inequalities and condition (2.5) we see that for
any $x \in \mathcal{X}$

$$\sum_{x' \in \mathcal{X}} |B(x'-x)| \leq B(0) + \frac{R}{A} + \frac{2}{A\pi^2} \sum_{j=1}^{\frac{R+1}{2}} \|j\delta\|^{-2}$$

$$< 2K + A + \frac{R}{A} + \frac{2\zeta(2)}{A\pi^2\delta^2}$$

$$= 2K + A + \frac{R}{A} + (3A\delta^2)^{-1}.$$

From (2.5) it is immediate that $\delta \leqslant R^{-1}$, and hence $R A^{-1} \leqslant (\delta A)^{-1}$. We take $A = [3^{-\frac{1}{2}} \delta^{-1}] + 1$, so that the above is

$$\leq 2K + 2 \cdot 3^{-\frac{1}{2}} \delta^{-1} + 1 + 3^{\frac{1}{2}}$$

$$\leq 2K + \frac{2}{\sqrt{9}} \delta^{-1} + 3.$$

This, with (2.19) implies (2.18), so the proof is complete.

Our brief proof of Theorem 2.3 is made possible by the sharp inequality (1.12), which gives immediate rise to (2.19). The earlier versions of the large sieve in which the best constant is attached to N were based on an observation of Davenport and Halberstam [47]. This principle states that under condition (2.5) we have

$$\sum_{x \in \mathcal{X}} |S(x)|^2 \leq \left(\sum |a_x|^2 c_n^{-2} \right) \left(\sum c_n^2 \right), \qquad (2.23)$$

where the c_n are real numbers such that $c_n \neq 0$ whenever $a_n \neq 0$, and

$$C(x) = \sum c(n) e(nx)$$

is square-summable and vanishes when $\|x\| > \frac{\delta}{2}$. The inequality (2.19) is easier to deal with than (2.23), because in (2.19) there is no condition that $B(x)$ vanish outside an interval. Nevertheless, it is interesting to note that we can derive (2.23) from (2.19). To do this we first note that if (2.5) holds and if $B(x)$ vanishes when $\|x\| > \delta$, then (2.19) may be written

$$\sum_{x \in \mathcal{X}} |S(x)|^2 \leq \left(\sum |a_n|^2 b_n^{-1} \right) \left(\sum b_n \right).$$

(In fact this follows from Bessel's inequality (1.9).) Following a well-known construction (see [196]) we note that if we take

$$B(x) = \int_0^1 C(t) \, C(x-t) \, dt$$

then $b_n = c_n^2$, and the condition that $C(x)$ vanish for $\|x\| > \frac{\delta}{2}$ implies that $B(x)$ vanishes for $\|x\| > \delta$. The C_n are real so the b_n are non-negative; hence (2.23) follows from (2.19).

To prove Theorem 2.4 we use (2.19), with a_k replaced by $a_k\left(1 - \frac{|k|}{K}\right)$; we take $b_k = \left(1 - \frac{|k|}{K}\right)$. Then

$$B(x) = K^{-1}\left(\frac{\sin \pi K x}{\sin \pi x}\right)^2 \le K^{-1}\left(\pi^{-2}\|x\|^{-2} + 1\right).$$

Hence for $x \in \mathfrak{X}$

$$\sum_{x' \in \mathfrak{X}} |B(x'-x)| \le B(0) + \frac{R}{K} + \frac{2}{K\pi^2} \sum_{j=1}^{\frac{R+1}{2}} \|j\delta\|^{-2}$$

$$\le K + \frac{R}{K} + 2\,\zeta(2)\,\pi^{-2}\,K^{-1}\,\delta^{-2}$$

$$\le K + (K\delta)^{-1} + (3K\delta^2)^{-1}$$

$$\le K\left(1 + (K\delta)^{-2}\right)$$

if $\delta \le \frac{1}{2}$. If $\delta > \frac{1}{2}$ then $\mathfrak{X}$ contains only one element. In this case we have from Cauchy's inequality

$$|S(x)|^2 \le \left(\sum \left(1 - \frac{|k|}{K}\right)\right)\left(\sum |a_k|^2\left(1 - \frac{|k|}{K}\right)\right)$$

$$= K \sum |a_k|^2\left(1 - \frac{|k|}{K}\right).$$

Thus (2.12) holds in either case.

We now give Gallagher's proof of Theorem 2.5. We first note that if $S(x)$ is defined by (2.2) then

$$\sum_{q \leq Q} \sum_{a=1}^{q} {}^{*} |S(\tfrac{a}{q})|^{2} \leq (\pi N + Q^{2}) \sum |a_{n}|^{2}. \tag{2.24}$$

In fact this follows from Corollary 2.2, and it suffices to show that $\delta \geq Q^{-2}$. If $\frac{a}{q}$ and $\frac{a'}{q'}$ are distinct modulo 1 and $q \leq Q$, $q' \leq Q$, then

$$\left\| \frac{a}{q} - \frac{a'}{q'} \right\| = \left\| \frac{a q' - a' q}{q q'} \right\| \geq \frac{1}{q q'} \geq \frac{1}{Q^{2}},$$

so $\delta \geq Q^{-2}$ as required.

It is known (see Davenport, §9) that

$$\chi(n) = \tau(\overline{\chi})^{-1} \sum_{b=1}^{q} {}^{*} \overline{\chi}(b) e(\tfrac{bn}{q}), \tag{2.25}$$

where $\tau(\chi)$ is the Gauss sum

$$\tau(\chi) = \sum_{b=1}^{q} {}^{*} \chi(b) e(\tfrac{b}{q}),$$

whenever at least one of

$$\chi \text{ is primitive (and } n \text{ is any integer)} \tag{2.26a}$$

or

$$(n,q) = 1 \quad (\text{and } \chi \text{ is any character mod } q) \tag{2.26b}$$

is true. (The relation (2.25) is used, under assumption (2.26a), in the proof of the functional equation for $L(s,\chi)$.) If χ is primitive then

$$T(\chi) \;=\; \sum a_n \chi(n)$$

$$=\; \tau(\bar{\chi})^{-1} \sum_{b=1}^{q}{}^{*}\, \bar{\chi}(b) \sum_{n} a_n e\!\left(\tfrac{bn}{q}\right)$$

$$=\; \tau(\bar{\chi})^{-1} \sum_{b=1}^{q}{}^{*}\, \bar{\chi}(b)\, S\!\left(\tfrac{b}{q}\right).$$

Now $|\tau(\chi)| = q^{\frac{1}{2}}$ for primitive χ, so

$$\sum_{\chi}{}^{*}\, |T(\chi)|^{2} \;=\; q^{-1} \sum_{\chi}{}^{*}\left|\sum_{b=1}^{q}{}^{*}\, \bar{\chi}(b)\, S\!\left(\tfrac{b}{q}\right)\right|^{2}$$

$$\leq\; q^{-1} \sum_{\chi}\left|\sum_{b=1}^{q}{}^{*}\, \bar{\chi}(b)\, S\!\left(\tfrac{b}{q}\right)\right|^{2}$$

$$=\; q^{-1}\phi(q) \sum_{b=1}^{q}{}^{*}\, \left|S\!\left(\tfrac{b}{q}\right)\right|^{2};$$

this and (2.24) imply (2.15).   If $a_n = 0$ whenever $(n, q) > 1$ then one can use (2.25) for any character, in virtue of (2.26b).   This led Bombieri and Davenport [20] to another result of this sort, inequality (3.3).

C H A P T E R   3

Arithmetic formulations of the large sieve

In its arithmetic setting, the large sieve is a statement concerning the distribution of a sequence in arithmetic progressions. Let $\mathcal{N}$ be a set of Z integers in the interval $[M+1, M+N]$, and put

$$Z(q, h) = \sum_{\substack{n \in \mathcal{N} \\ n \equiv h(q)}} 1.$$

If we set

$$S(x) = \sum_{n \in \mathcal{N}} e(nx),$$

then it is easily proved that for any prime p

$$p \sum_{h=1}^{p} \left(Z(p, h) - \frac{Z}{p} \right)^2 = \sum_{a=1}^{p-1} \left| S\left(\frac{a}{p}\right) \right|^2. \tag{3.1}$$

Hence from Theorem 2.3 we see that

$$\sum_{p \leq X} p \sum_{h=1}^{p} \left(Z(p, h) - \frac{Z}{p} \right)^2 \leq \left(N + \frac{2}{\sqrt{3}} X^2 + 3 \right) Z. \tag{3.2}$$

The expression on the left was implicit in the original work of Linnik [122], and was later treated explicitly by Rényi [155], [156]. The bound (3.2) is near to being best possible, though Wolke has shown (in a paper to appear) that in some cases slightly more is true.

Bombieri and Davenport [20] were first to formulate the

large sieve for purposes of a "small sieve". Their result is
as follows: Let P be a set of primes p for which $a_n = 0$
whenever $p | n$. Let $\mathcal{Q}$ be those integers composed solely of the
primes in P. Then

$$\sum_{\substack{q \leq Q \\ q \in \mathcal{Q}}} \phi(q)^{-1} \sum_{\chi} |\tau(\chi)|^2 \left| \sum_{M+1}^{M+N} a_n \chi(n) \right|^2 \leq \left(N + \frac{2}{\sqrt{3}} Q^2 + 3 \right) \sum_{M+1}^{M+N} |a_n|^2. \tag{3.3}$$

The contribution of the principal characters on the left is

$$= \left(\sum_{\substack{q \leq Q \\ q \in \mathcal{Q}}} \frac{\mu^2(q)}{\phi(q)} \right) \left| \sum a_n \right|^2,$$

and the inequality

$$\left(\sum_{\substack{q \leq Q \\ q \in \mathcal{Q}}} \frac{\mu^2(q)}{\phi(q)} \right) \left| \sum a_n \right|^2 \leq \left(N + \frac{2}{\sqrt{3}} Q^2 + 3 \right) \sum |a_n|^2$$

can be used to obtain small sieve results in which at most
one residue class is removed.

It would seem to require considerable ingenuity to extend
(3.3) to a "nice" result which is suitable for sieving more
residue classes. On the other hand the advantage of (3.3)
over (3.2) is seen to lie in the inclusion of composite q
in the sum; this suggests that one might obtain sharp results
if one could extend (3.1) to composite moduli. This can be
done (see [141]); the result is that

$$q \sum_{h=1}^{q} \left| \sum_{d | q} \frac{\mu(d)}{d} Z\left(\frac{q}{d}, h\right) \right|^2 = \sum_{a=1}^{q} {}^* \left| S\left(\frac{a}{q}\right) \right|^2. \tag{3.4}$$

Schaal [183] has recently extended this to algebraic number fields.   Using (3.4) we may prove

THEOREM 3.1.   Let a_n $(M+1 \le n \le M+N)$ be arbitrary real or complex numbers.   For each prime p let $\omega(p)$ be the number of residue classes $h \bmod p$ for which $a_n = 0$ whenever $n \equiv h \pmod{p}$.   Then for any $Q \ge 1$

$$\left| \sum_n a_n \right|^2 \le \frac{N + 2Q^2}{L} \sum_n |a_n|^2,$$ (3.5)

where

$$L = \sum_{q \le Q} \mu^2(q) \prod_{p | q} \frac{\omega(p)}{p - \omega(p)}.$$ (3.6)

We note that the result is true but uninteresting in the two extreme cases $\omega(p) = p$ for some $p \le Q$ and $\omega(p) = 0$ for all $p \le Q$.   In the former case $L = \infty$, and in the latter $L = 1$.

If the a_n take only the values 0 and 1 then $\sum |a_n|^2 = \sum a_n$, so we have

COROLLARY 3.2.   Let $\mathcal{N}$ be a set of Z integers in the interval $[M+1 , M+N]$.   For prime p let $\omega(p)$ denote the number of residue classes $\bmod p$ that contain no element of $\mathcal{N}$. Then for $Q \ge 1$

$$Z \le \frac{N + 2Q^2}{L},$$ (3.7)

where L is defined by (3.6).

Selberg's upper bound method [192], [193], [194] gives a surprisingly similar bound:

$$Z \leq \frac{N}{L} + R. \qquad (3.8)$$

Here R is an error term which must be majorized. When the $\omega(p)$ are small a satisfactory majorization can be given if Q is somewhat smaller than $N^{\frac{1}{2}}$. The advantage of (3.7) is that it holds regardless of the size of the $\omega(p)$. Moreover, (3.8) has the disadvantage that further work is required (in estimating R) before it becomes useful. Nevertheless, Selberg's upper bound method has the distinct advantage of flexibility and generality, particularly when applied to weighted and non-linear sets n (see [76], [163], [164]).

While (3.7) gives only an upper bound for Z, it is known that lower bounds may be deduced from upper bounds, by means of a combinatorial identity of Buchstab.

When the $\omega(p)$ are so large that $p - \omega(p)$ is small on average, the elementary sieve of Gallagher [70] gives stronger results than (3.7). In addition to the small sieves and the sieve of Gallagher, one should keep in mind the function-theoretic sieve methods of Turán [215], [216], [217], [218], and of Golomb [72].

Following my work Wirsing, Richert, Gallagher, and Huxley [94] have each given other proofs of Theorem 3.1. In addition, Huxley [91] and Wilson have generalized (3.2) to algebraic number fields, and Huxley [91] has also generalized (3.3) and (3.5) to algebraic number fields. Johnsen has proved a large sieve and an analogue of (3.5) in rational function fields in one variable over a finite constant field.

The central feature of the proof of Theorem 3.1 is the inequality

$$\left| \sum_n a_n \right|^2 \mu^2(q) \prod_{p|q} \frac{\omega(p)}{p - \omega(p)} \leq \sum_{a=1}^{q}{}^* |S(\tfrac{a}{q})|^2. \tag{3.9}$$

My original proof [141] of this was based on the identity (3.4), which meant that (3.4) also required proof. We now give Huxley's [94] simple and direct proof of (3.9).

If q has a squared factor then (3.9) is trivial, so we assume that q is square-free. Let $R(q)$ consist of those numbers $r, 1 \leq r \leq q$, for which $(n-r, q) = 1$ whenever $a_n \neq 0$. From the definition of the $\omega(p)$ and the Chinese remainder theorem we see that $R(q)$ contains precisely $\prod_{p|q} \omega(p)$ numbers. We use the Jensen - Ramanujan identity (see Hardy and Wright, §16.6) for the Möbius function:

$$\mu(q) = \sum_{a=1}^{q}{}^* e(\tfrac{a}{q}).$$

From this it is clear that if $(h, q) = 1$ then

$$\mu(q) = \sum_{a=1}^{q}{}^* e(\tfrac{ah}{q}). \tag{3.10}$$

Hence for any $r \in \mathcal{R}(q)$ we have

$$a_n \mu(q) = \sum_{a=1}^{q} {}^* a_n \, e\left(\frac{(n-r)a}{q}\right).$$

We sum this over n and all $r \in \mathcal{R}(q)$ to obtain

$$\left(\sum_n a_n\right) \mu(q) \prod_{p|q} \omega(p) = \sum_{a=1}^{q} {}^* \left(\sum_n a_n e\left(\frac{an}{q}\right)\right)\left(\sum_{r \in \mathcal{R}(q)} e\left(\frac{-ar}{q}\right)\right)$$

$$= \sum_{a=1}^{q} {}^* S\left(\frac{a}{q}\right) \sum_{r \in \mathcal{R}(q)} e\left(\frac{-ar}{q}\right).$$

By Cauchy's inequality we see that

$$\left|\sum a_n\right|^2 \mu^2(q) \prod_{p|q} \omega(p)^2 \le \left(\sum_{a=1}^{q} {}^* \left|S\left(\frac{a}{q}\right)\right|^2\right)\left(\sum_{a=1}^{q} {}^* \left|\sum_{r \in \mathcal{R}(q)} e\left(\frac{-ar}{q}\right)\right|^2\right). \qquad (3.11)$$

Now a little thought discloses that for square-free q the second factor is multiplicative. Hence it is

$$= \prod_{p|q}\left(\sum_{a=1}^{p-1} \left|\sum_{r \in \mathcal{R}(p)} e\left(\frac{-ar}{p}\right)\right|^2\right)$$

$$= \prod_{p|q}\left(\sum_{r_1 \in \mathcal{R}(p)} \sum_{r_2 \in \mathcal{R}(p)} \sum_{a=1}^{p-1} e\left(\frac{a(r_2-r_1)}{p}\right)\right).$$

The value of the innermost sum is $p-1$ or -1 according as $p|(r_2-r_1)$ or not, so the above is

$$= \prod_{p|q} \left((p-1)\,\omega(p) - \omega(p)(\omega(p)-1) \right)$$

$$= \prod_{p|q} \omega(p)(p - \omega(p)).$$

This, with (3.11), proves (3.9).

We now derive Theorem 3.1. We take $\mathfrak{X}$ in Theorem 2.3 to be the fractions $\frac{a}{q}$, $q \leqslant Q$, and we recall from the proof of Theorem 2.5 that in this case $\delta \geqslant Q^{-2}$. If $Q \geq 2$ then $N + \frac{2}{\sqrt{3}} Q^2 + 3 \leqslant N + 2 Q^2$, so we have

$$\sum_{q \leqslant Q} \sum_{a=1}^{q} {}^* \left| S\left(\tfrac{a}{q}\right) \right|^2 \leqslant (N + 2Q^2) \sum |a_n|^2. \tag{3.12}$$

If $Q < 2$ then the above remains valid, in view of (2.9). The above and (3.9) give (3.5), so the proof is complete.

One can extend Theorem 3.1 by stating a result which allows one to sieve by higher powers of primes; Johnsen has produced such a result. One can proceed as above, but using the identities of Cohen [35] in place of (3.10). In this manner one can give an upper bound for the number of square-free numbers in an interval, though the result is inferior to that which Richert (unpublished) has obtained using Selberg's method.

We note that our remarks in the previous chapter do not imply that (3.12) is best possible. Erdős [58] and Erdős and Rényi [59] have investigated the possibility of improving on the large sieve when $\mathfrak{X}$ is taken to be the Farey fractions, as in (3.12).   They found (see also Elliott [55]) that if $S(x)$ is arbitrary, given by (2.2), then

$$\sum_{q \leq Q} \sum_{a=1}^{q} {}^{*} |S(\tfrac{a}{q})|^2 \ll N \sum |a_n|^2$$

only if $Q \ll N^{\frac{1}{2}}$.

A weighted sieve and its application

Although the large sieve in the forms of inequalities
(2.7) and (2.10) is essentially best possible, it nevertheless may be the case that if $S(x)$ is given by (2.2) then

$$\sum_{x \in \mathcal{X}} (N + C \delta(x)^{-1})^{-1} |S(x)|^2 \leq \sum_{n} |a_n|^2, \tag{4.1}$$

where

$$\delta(x) = \min_{\substack{x' \in \mathcal{X} \\ x' \neq x}} \|x - x'\|, \tag{4.2}$$

and $C > 0$ is some absolute constant. This would represent
a considerable sharpening of (2.10) when the $x \in \mathcal{X}$ are
irregularly spaced. This was the case in the application
giving (3.12); we could replace (3.12) by

$$\sum_{q \leq Q} \sum_{a=1}^{q} {}^{*} (N + C_1 q)^{-1} |S(\tfrac{a}{q})|^2 \leq \sum_{n} |a_n|^2, \tag{4.3}$$

by using (4.1) instead of (2.10). The consequent modification
of Theorem 3.1 would lead to improvements of several important
bounds. Although I have been unsuccessful in my attempts to
prove (4.1) and (4.3), the following more complicated result is
sufficient for arithmetic purposes.

THEOREM 4.1. <u>Let</u> a_n $(M+1 \leq n \leq M+N)$ <u>be arbitrary</u>
<u>real or complex numbers. There exist real numbers</u>
$b_n = b_n (M, N; Q, p)$ <u>for which</u> $b_n \geq 1$ <u>for</u> $M+1 \leq n \leq M+N,$

and

$$\sum_{q \leq Q} \left(N + \tfrac{15}{8} q Q\right)^{-1} \sum_{a=1}^{q} {}^{*} \left| \sum_{n=M+1}^{M+N} a_n b_n e\left(\tfrac{a n}{q}\right)\right|^2 \leq \sum_{n} |a_n|^2. \qquad (4.4)$$

This sharpens a result of Bombieri [16], who obtained (4.4) with the factor $\left(N + C\left(N_q Q\right)^{\frac{1}{2}}\right)$ in place of $\left(N + \tfrac{15}{8} q Q\right)$. From Theorem 4.1 we deduce

THEOREM 4.2. Let a_n $(M+1 \leq n \leq M+N)$ be arbitrary real or complex numbers. For each prime p let $w(p)$ be the number of residue classes $h \bmod p$ for which $a_n = 0$ whenever $n \equiv h \pmod p$. Then for $Q \geq 1$

$$\left| \sum_{n} a_n \right|^2 \leq \frac{N}{L'} \sum_{n} |a_n|^2, \qquad (4.5)$$

where

$$L' = \sum_{q \leq Q} \mu^2(q) \left(1 + \tfrac{15}{8} q Q N^{-1}\right)^{-1} \prod_{p \mid q} \frac{w(p)}{p - w(p)}. \qquad (4.6)$$

From this we have

COROLLARY 4.3. Let n be a set of Z integers in the interval $[M+1, M+N]$. For prime p let $w(p)$ denote the number of residue classes $\bmod p$ that contain no element of n. Then for $Q \geq 1$

$$Z \leq \frac{N}{L'}, \qquad (4.7)$$

where L' is defined by (4.6)

Using Selberg's method (see [102]) one may show that

$$\pi(M+N; q, a) - \pi(M; q, a) \leq \frac{2N}{\phi(q) \log \frac{N}{q}} \left(1 + O\left(\frac{\log\log \frac{N}{q}}{\log \frac{N}{q}}\right)\right) \tag{4.8}$$

when $q \leq \frac{1}{3} N$.   Bombieri and Davenport [20] gave a second proof of this; the inequality (3.7) would serve equally well in this connection.   By obtaining a better estimate for R in (3.8), van Lint and Richert [134] sharpened (4.8) to

$$\pi(M+N; q, a) - \pi(M; q, a) \leq \frac{2N}{\phi(q) \log \frac{N}{q}} \left(1 + O\left(\frac{1}{\log \frac{N}{q}}\right)\right) \tag{4.9}$$

when $q \leq \frac{1}{3} N$.   More recently Bombieri [16] used his weaker form of (4.4) to give a second proof of (4.9).   In particular he showed that

$$\pi(M+N) - \pi(M) \leq \frac{2N}{\log N - 3} .$$

We obtain an even sharper result from the bound (4.7).

THEOREM 4.4.   There exists a positive absolute constant $\delta > 0$ such that if $q < \delta N$ then

$$\pi(M+N; q, a) - \pi(M; q, a) \leq \frac{2N}{\phi(q) \log \frac{N}{q}} . \tag{4.10}$$

When $q = 1$ the condition $\delta N > 1$ means merely that N is sufficiently large.   Recently Vaughan has studied the question of the admissible size of δ .   He has found that one may take $\delta = 1$, so in particular

$$\pi(M+N) - \pi(M) \leqslant \frac{2N}{\log N} \qquad (4.11)$$

for any $N > 1$. Of course this still falls far short of the long-standing conjecture that

$$\pi(M+N) \leqslant \pi(M) + \pi(N) \qquad (4.12)$$

when $M > 1$, $N > 1$. We know that this conjecture is true [184] , [185] when $1 < \min(M,N) \leqslant 146$, or [104, §58] when $M = N \geqslant N_0$. As (4.12) is thus true for $N \leqslant 17$, and [175] as

$$\frac{N}{\log N} \leqslant \pi(N)$$

for $N \geq 17$, we see that (4.11) implies that

$$\pi(M+N) - \pi(M) \leqslant 2\pi(N), \qquad (4.13)$$

for positive integers M and $N > 1$. However, this is rather weaker than (4.11) for large N , as

$$\pi(N) = \frac{N}{\log N - 1} + O\left(\frac{N}{(\log N)^3}\right).$$

In proving Theorem 4.1 we require the following technical lemma, which elaborates on the work of Bombieri and Davenport [21] .

LEMMA 4.5. Let δ satisfy $0 < \delta \leqslant \frac{1}{2}$, and let N be a positive integer. Let $f_i(x) = f_i(x, N, \delta)$ $(i = 1, 2, 3, 4)$ be real functions of the real variable x , possessing periods 1 , which in $[-\frac{1}{2}, \frac{1}{2}]$ are defined by the relations

$$f_1(x) = f_2(x)\big(f_3(x) + f_4(x)\big), \tag{4.14}$$

$$f_2(x) = \begin{cases} \cos \dfrac{\pi x}{2\delta} & |x| \le \delta, \\[2mm] 0 & \delta < |x| \le \tfrac{1}{2}, \end{cases} \tag{4.15}$$

$$f_3(x) = \frac{\sin \pi\left(2N + 2^{-\frac{1}{2}}\delta^{-1}\right)x}{\pi x}, \tag{4.16}$$

and

$$f_4(x) = 2^{-1}\delta^{-2}(\cos 2\pi Nx)\max(0, \delta - |x|). \tag{4.17}$$

Let $b_n = b_n(N, \delta)$ be defined by the relation

$$f_1(x) = \sum_{-\infty}^{+\infty} b_n e(nx). \tag{4.18}$$

Then

$$b_n \ge 1 \qquad (|n| \le N), \tag{4.19}$$

and

$$\int_{-\delta}^{\delta} \big(f_3(x) + f_4(x)\big)^2 \, dx \le 2N + \tfrac{15}{8}\delta^{-1}. \tag{4.20}$$

To establish (4.19) it suffices to show that

$$\int_0^1 e(nx) f_2(x) f_3(x) \ge 1 - \frac{1}{8\pi^2\left((N-n)\delta + \frac{1}{4}\right)^2} - \frac{1}{8\pi^2\left((N+n)\delta + \frac{1}{4}\right)^2} \tag{4.21}$$

and

$$\int_0^1 e(nx) f_2(x) f_4(x)\,dx \geqslant \frac{1}{8\pi^2((N-n)\delta + \frac{1}{4})^2} + \frac{1}{8\pi^2((N+n)\delta + \frac{1}{4})^2} \tag{4.22}$$

when $|n| \leq N$. We first prove (4.21). We write the left hand side of (4.21) as

$$\frac{1}{4\pi i}\left(F(N\delta + n\delta + 2^{-\frac{3}{2}} + \tfrac{1}{4}) + F(N\delta + n\delta + 2^{-\frac{3}{2}} - \tfrac{1}{4})\right. \tag{4.23}$$
$$\left. - F(-N\delta + n\delta - 2^{-\frac{3}{2}} - \tfrac{1}{4}) - F(-N\delta + n\delta - 2^{-\frac{3}{2}} + \tfrac{1}{4})\right),$$

where

$$F(\alpha) = \int_{-1}^1 \frac{e(\alpha t)}{t}\,dt,$$

and the Cauchy principal value is taken at 0 . Now

$$F(\alpha) = \pi i\,\mathrm{sgn}\,\alpha - \int_{-\infty}^{-1} \frac{e(\alpha t)}{t}\,dt - \int_1^\infty \frac{e(\alpha t)}{t}\,dt,$$

so for $\alpha > 0$

$$F(\alpha) + F(\alpha + \tfrac{1}{2}) = -F(-\alpha) - F(-\alpha - \tfrac{1}{2})$$
$$= 2\pi i - \int_1^\infty \frac{e(\alpha t) + e((\alpha + \frac{1}{2})t)}{t}\,dt - \int_{-\infty}^{-1} \frac{e(\alpha t) + e((\alpha + \frac{1}{2})t)}{t}\,dt.$$

These integrals are of the same modulus, so it suffices to consider the first one. On rotating the path of integration through an angle of $\frac{\pi}{2}$ we see by Cauchy's theorem that

$$\left| \int_1^\infty \frac{e(\alpha t) + e((\alpha + \frac{1}{2})t)}{t} \, dt \right| = \left| \int_0^\infty \frac{e^{-2\pi\alpha u} - e^{-2\pi(\alpha + \frac{1}{2})u}}{1 + iu} \, du \right|$$

$$\leq \int_0^\infty \left| e^{-2\pi\alpha u} - e^{-2\pi(\alpha + \frac{1}{2})u} \right| \, du$$

$$= \frac{1}{2\pi\alpha} - \frac{1}{2\pi(\alpha + \frac{1}{2})}$$

$$= \frac{1}{2\pi\alpha(2\alpha + 1)} .$$

From (4.23) we see that the left hand side of (4.21) is

$$\geq 1 - \frac{1}{8\pi^2 \left((N-n)\delta + 2^{-\frac{1}{2}} - \frac{1}{4} \right) \left((N-n)\delta + 2^{-\frac{1}{2}} + \frac{1}{4} \right)}$$

$$- \frac{1}{8\pi^2 \left((N+n)\delta + 2^{-\frac{1}{2}} - \frac{1}{4} \right) \left((N+n)\delta + 2^{-\frac{1}{2}} + \frac{1}{4} \right)}$$

$$\geq 1 - \frac{1}{8\pi^2 \left((N-n)\delta + \frac{1}{4} \right)^2} - \frac{1}{8\pi^2 \left((N+n)\delta + \frac{1}{4} \right)^2}$$

when $|n| \leq N$.

To prove (4.22) we note that the left hand side may be written

$$\frac{1}{8} \left(G\left((N-n)\delta + \frac{1}{4} \right) + G\left((N-n)\delta - \frac{1}{4} \right) + G\left((N+n)\delta + \frac{1}{4} \right) + G\left((N+n)\delta - \frac{1}{4} \right) \right), \tag{4.24}$$

where

$$G(\alpha) = \int_{-1}^{1} e(\alpha x)(1-|x|)\,dx$$

$$= \left(\frac{\sin \pi \alpha}{\pi \alpha}\right)^{2}.$$

Hence for $\alpha \geq 0$

$$G(\alpha + \tfrac{1}{4}) + G(\alpha - \tfrac{1}{4}) = \frac{(\sin \pi(\alpha + \tfrac{1}{4}))^{2}}{\pi^{2}(\alpha + \tfrac{1}{4})^{2}} + \frac{(\cos \pi(\alpha + \tfrac{1}{4}))^{2}}{\pi^{2}(\alpha - \tfrac{1}{4})^{2}}$$

$$\geq \frac{1}{\pi^{2}(\alpha + \tfrac{1}{4})^{2}}.$$

Thus from (4.24) the left hand side of (4.22) is

$$\geq \frac{1}{8\pi^{2}((N-n)\delta + \tfrac{1}{4})^{2}} + \frac{1}{8\pi^{2}((N+n)\delta + \tfrac{1}{4})^{2}}.$$

This completes the proof of (4.19)

To obtain (4.20) we expand the square in the integrand
and estimate each integral separately.   We have

$$\int_{-\delta}^{\delta} (f_{3}(x))^{2}\,dx \leq \pi^{-2}\int_{-\infty}^{\infty} \left(\frac{\sin(2N + 2^{-\frac{1}{2}}\delta^{-1})\pi x}{x}\right)^{2}\,dx$$

$$= 2N + 2^{-\frac{1}{2}}\delta^{-1}. \tag{4.25}$$

Also

$$2 \int_{-\delta}^{\delta} f_3(x) f_4(x) dx = 2^{-1} \pi^{-1} \delta^{-2} \int_{-\delta}^{\delta} \left(\sin (4N + 2^{-\frac{1}{2}} \delta^{-1}) \pi x + \sin 2^{-\frac{1}{2}} \delta^{-1} \pi x \right) \cdot$$

$$\cdot (\delta - |x|) \frac{dx}{x}.$$

But for $\alpha > 0$

$$\int_{-1}^{1} (\sin 2\pi \alpha x)(1 - |x|) \frac{dx}{x} = \pi - 2 \int_{1}^{\infty} (\sin 2\pi \alpha x) \frac{dx}{x} - \frac{1 - \cos 2\pi \alpha}{\pi \alpha}$$

$$= \pi - \frac{2}{\pi \alpha} \int_{1}^{\infty} \left(\frac{\sin \pi \alpha x}{x} \right)^2 dx$$

$$\leq \pi,$$

so

$$2 \int_{-\delta}^{\delta} f_3(x) f_4(x) dx \leq \delta^{-1}. \tag{4.26}$$

Finally,

$$\int_{-\delta}^{\delta} (f_4(x))^2 dx \leq 2^{-2} \delta^{-4} \int_{-\delta}^{\delta} (\delta - |x|)^2 dx$$

$$= (6\delta)^{-1}. \tag{4.27}$$

Adding (4.25), (4.26), (4.27), we have (4.20), since

$$2^{-\frac{1}{2}} + 1 + \frac{1}{6} = 1.873773\ldots < \frac{15}{8}.$$

This completes the proof of Lemma 4.5.

We now prove Theorem 4.1. As in the proofs of Theorems

2.1 and 2.3, it suffices to show that if

$$S(x) = \sum_{-N}^{N} a_n e(nx) \tag{4.28}$$

then

$$\sum_{q \leq Q} \left(2N + \tfrac{15}{8} q Q\right)^{-1} \sum_{a=1}^{q} {}^{*} \left| \sum_{-N}^{N} a_n b_n e\left(\tfrac{an}{q}\right) \right|^2 \leq \sum_{-N}^{N} |a_n|^2, \tag{4.29}$$

where $b_n = b_n(N; Q, q) \geq 1$ for $|n| \leq N$.

Let $b_n = b_n(N; Q, q)$ be the coefficients in (4.18) with $\delta = (qQ)^{-1}$. Then

$$\sum_{-N}^{N} a_n b_n e\left(\tfrac{an}{q}\right) = \int_0^1 S\left(t + \tfrac{a}{q}\right) f_1(-t)\,dt = \int_{\tfrac{a}{q} - \tfrac{1}{qQ}}^{\tfrac{a}{q} + \tfrac{1}{qQ}} S(u) f_1\left(\tfrac{a}{q} - u\right)du,$$

so by the Cauchy - Schwarz inequality

$$\left| \sum_{-N}^{N} a_n b_n e\left(\tfrac{an}{q}\right) \right|^2 \leq \int_{\tfrac{a}{q} - \tfrac{1}{qQ}}^{\tfrac{a}{q} + \tfrac{1}{qQ}} |S(u) f_2(u - \tfrac{a}{q})|^2 du \int_{-\tfrac{1}{qQ}}^{\tfrac{1}{qQ}} \left(f_3(u) + f_4(u)\right)^2 du.$$

From (4.20) we have

$$\left(2N + \tfrac{15}{8} q Q\right)^{-1} \left| \sum_{-N}^{N} a_n b_n e\left(\tfrac{an}{q}\right) \right|^2 \leq \int_{\tfrac{a}{q} - \tfrac{1}{qQ}}^{\tfrac{a}{q} + \tfrac{1}{qQ}} |S(u) f_2(u - \tfrac{a}{q})|^2 du.$$

If $\tfrac{a}{q}$ and $\tfrac{a'}{q'}$ are neighboring Farey fractions then for u between $\tfrac{a}{q}$ and $\tfrac{a'}{q'}$

$$f_2\left(u-\tfrac{a}{q}, \tfrac{1}{qQ}\right)^2 + f_2\left(u-\tfrac{a'}{q'}, \tfrac{1}{q'Q}\right)^2$$

$$\le f_2\left(u-\tfrac{a}{q}, \tfrac{1}{qq'}\right)^2 + f_2\left(u-\tfrac{a'}{q'}, \tfrac{1}{qq'}\right)^2$$

$$= 1,$$

so

$$\sum_{q\le Q}\left(2N+\tfrac{15}{8}qQ\right)^{-1}\sum_{a=1}^{q}{}^{*}\left|\sum_{-N}^{N} a_n b_n e\left(\tfrac{an}{q}\right)\right|^2 \le \int_0^1 |S(u)|^2 du$$

$$= \sum_{-N}^{N} |a_n|^2,$$

which is (4.29). The condition on the b_n is given by (4.19).

We now prove Theorem 4.2. If each a_n is replaced by $|a_n|$ then the left hand side of (4.5) is not decreased, while the right hand side remains unchanged. Hence it suffices to prove (4.5) when the a_n are non-negative real numbers. From (3.9) and Theorem 4.1 we have immediately

$$\sum_{q\le Q}\mu^2(q)\left(N+\tfrac{15}{8}qQ\right)^{-1}\left(\prod_{p|q}\tfrac{\omega(p)}{p-\omega(p)}\right)\left|\sum_{n} a_n\right|^2$$

$$\le \sum_{q\le Q}\mu^2(q)\left(N+\tfrac{15}{8}qQ\right)^{-1}\left(\prod_{p|q}\tfrac{\omega(p)}{p-\omega(p)}\right)\left|\sum_{n} a_n b_n\right|^2$$

$$\le \sum_{q\le Q}\left(N+\tfrac{15}{8}qQ\right)^{-1}\sum_{a=1}^{q}{}^{*}\left|\sum_{-N}^{N} a_n b_n e\left(\tfrac{an}{q}\right)\right|^2$$

$$\le \sum_{-N}^{N} |a_n|^2,$$

so (4.5) follows.

Corollary 4.3 follows from Theorem 4.2, on taking $a_n = 1$ or 0 according as $n \in \mathcal{N}$ or not.

To derive Theorem 4.4 we let $\mathcal{N}$ be the set of those n for which $nq+a$ is prime, $M+1 \leq nq+a \leq M+N$, and $Q < nq+a$. These n lie in an interval of length

$$K = \left[\frac{M+N-a}{q} \right] - \left[\frac{M+1-a}{q} \right] < \frac{N}{q} + 1,$$

and $\omega(p) \geq 1$ if $p \nmid q$, $p \leq Q$. Hence

$$\pi(M+N; q, a) - \pi(M; q, a) \leq \frac{K}{L'} + \pi(Q; q, a)$$

$$\leq \frac{K}{L'} + Q. \tag{4.30}$$

We now require a lower bound for L' . We have

$$L' \geq \sum_{\substack{m \leq Q \\ (m,q)=1}} \mu^2(m)\, \phi(m)^{-1} \left(1 + \tfrac{15}{8} m Q K^{-1}\right)^{-1},$$

and

$$\frac{q}{\phi(q)} \sum_{\substack{m \leq Q \\ (m,q)=1}} \mu^2(m)\, \phi(m)^{-1} \left(1 + \tfrac{15}{8} m Q K^{-1}\right)^{-1} = \prod_{p \mid q} \left(1 + \frac{1}{p-1}\right) \sum_{\substack{m \leq Q \\ (m,q)=1}} \frac{\mu^2(m)}{\phi(m)} \left(1 + \tfrac{15}{8} m Q K^{-1}\right)^{-1}$$

$$\geq \sum_{m \leq Q} \frac{\mu^2(m)}{\phi(m)} \left(1 + \tfrac{15}{8} m Q K^{-1}\right)^{-1}.$$

It is known (see [133]) that if we define $E(x)$ by the relation

$$\sum_{m \leq x} \frac{\mu^2(m)}{\phi(m)} = \log x + \gamma + \sum_{p} \frac{\log p}{p(p-1)} + E(x),$$

then $E(x) = o(1)$ as $x \to \infty$. Hence by partial summation if $Q \leq K^{\frac{1}{2}}$ then

$$\sum_{m \leq Q} \mu^2(m)\, \phi(m)^{-1}\left(1 + \tfrac{15}{8} m Q K^{-1}\right)^{-1} = \log Q + \gamma + \sum_{p} \frac{\log p}{p(p-1)}$$

$$- \log\left(1 + \tfrac{15}{8} Q^2 K^{-1}\right) + o(1),$$

so on taking $Q = \left[\left(\tfrac{8}{15} K\right)^{\frac{1}{2}}\right]$ this is

$$= \tfrac{1}{2} \log K + \gamma + \sum_{p} \frac{\log p}{p(p-1)} - \tfrac{1}{2} \log \tfrac{15}{8} - \log 2 + o(1).$$

But

$$\log 2 = 0.69315\ldots,$$
$$\tfrac{1}{2} \log \tfrac{15}{8} = 0.31430\ldots,$$

and equations (2.8) and (2.11) of [175] give

$$\gamma + \sum_{p} \frac{\log p}{p(p-1)} = 1.332582\ldots,$$

so we have in all

$$L' \geq \frac{\phi(q)}{q}\left(\tfrac{1}{2} \log K + \tfrac{3}{10}\right)$$

provided $q < \delta_1 N$. This, with (4.30), gives

$$\pi(M+N;q,a) - \pi(M;q,a) \leq \frac{2Kq}{\phi(q)\left(\log K + \frac{6}{10}\right)} + K^{\frac{1}{2}}$$

$$\leq \frac{2N}{\phi(q)\log\frac{N}{q}}$$

when $\frac{N}{q}$ is sufficiently large.

A lower bound of Roth

Throughout our discussion of the large sieve we ignored the question of whether our ideas could be used to obtain lower bounds.   In this direction Roth [176] has given a lower bound which corresponds to the upper bound (3.2).   Thus Roth's result implies that in general a sequence cannot be too well distributed in all arithmetic progressions.   Huxley [94] has extended this result to sequences which have been sifted;   he finds that a sifted sequence cannot be too evenly distributed in those arithmetic progressions in which it lies.   In addition Roth [178], [180] and Choi [34] have extended the original result in other directions.   We confine ourselves to Roth's first result;   this we sharpen slightly.   Our proof differs substantially from Roth's;   we draw on ideas that we have used in previous chapters.

THEOREM 5.1.   Let η be a set of Z numbers in the interval $[1, N]$.   Let

$$Z(m; q, h) = \sum_{\substack{n \in \eta \\ n \leq m \\ n \equiv h \,(q)}} 1 , \tag{5.1}$$

let

$$\rho = \frac{Z}{N} , \tag{5.2}$$

and let

$$V_q(m) = \sum_{h=0}^{q-1} \left(Z(m; q, h) - \rho \left[\frac{m-h}{q} \right] \right)^2 . \tag{5.3}$$

<u>Then</u>

$$\rho^{(1-\rho)} N Q^2 \ll Q \sum_{q \leq Q} V_q(N) \left(\log \frac{2Q}{q} \right)^{-1} + \sum_{q \leq Q} q^{-1} \sum_{n=1}^{N} V_q(m). \qquad (5.4)$$

Roth obtained this, but without the factor $\left(\log \frac{2Q}{q} \right)^{-1}$ on the right hand side. We note that the lower bound (5.4) vanishes when $\rho = 0$ or $\rho = 1$. This is necessary, for in these cases the right hand side of (5.4) also vanishes. One might like to say that

$$\rho^{(1-\rho)} N Q^2 \ll Q \sum_{q \leq Q} V_q(N) \left(\log \frac{2Q}{q} \right)^{-1},$$

but this is false in general. For if $\mathcal{N}$ consists of the integers in $[1, \frac{N}{2}]$, then $\rho \sim \frac{1}{2}$, while $V_q(N) \ll q$ so the right hand side is $\ll Q^3$.

From (5.4) we have immediately

COROLLARY 5.2.   <u>In the above notation, there exist</u> $m \leq N$, $q \leq N^{\frac{1}{2}}$, $h \bmod q$, <u>for which</u>

$$\left| Z(m; q, h) - \rho \left[\frac{m-h}{q} \right] \right| \gg \left(\frac{\rho(1-\rho)N}{q \log N} \right)^{\frac{1}{2}}. \qquad (5.5)$$

We now prove Theorem 5.1.   We let

$$S(\alpha, u) = \sum_{\substack{n \in \mathcal{N} \\ n \leq u}} e(n\alpha) - \rho \sum_{1 \leq m \leq u} e(m u),$$

and we put $S(\alpha) = S(\alpha, N)$. Recalling the Farey dissection we see that the intervals $\mathfrak{m}_{\frac{a}{q}} = [\frac{a}{q} - \frac{1}{qQ}, \frac{a}{q} + \frac{1}{qQ}]$, $1 \le a \le q$, $(a, q) = 1$, $q \le Q$, cover the unit interval. Hence by Parseval's identity

$$\rho(1-\rho)N = \int_0^1 |S(\alpha)|^2 d\alpha$$

$$\le \sum_{q \le Q} \sum_{a=1}^q {}^* \int_{\mathfrak{m}_{\frac{a}{q}}} |S(\alpha)|^2 d\alpha. \qquad (5.6)$$

We now must obtain an estimate for $\int_{\mathfrak{m}_{\frac{a}{q}}} |S(\alpha)|^2 d\alpha$ in terms of $|S(\frac{a}{q})|^2$. By partial summation (this is equation (151) in Estermann's tract)

$$S(\tfrac{a}{q} + \beta) = e(N\beta) S(\tfrac{a}{q}) - 2\pi i \beta \int_0^N e(u\beta) S(\tfrac{a}{q}, u) du.$$

Hence

$$\int_{-\delta}^{\delta} |S(\tfrac{a}{q} + \beta)|^2 d\beta \ll \delta |S(\tfrac{a}{q})|^2 + \delta^2 \int_{-\delta}^{\delta} \left| \int_0^N e(u\beta) S(\tfrac{a}{q}, u) du \right|^2 d\beta. \qquad (5.7)$$

To estimate the second term we observe that it is bilinear in $S(\frac{a}{q}, u)$, and we draw on an idea in the proof of Lemma 1.6. For an arbitrary continuous function $f(u)$ we have

$$\int_{-\delta}^{\delta} \left| \int_0^N e(u\beta) f(u) du \right|^2 d\beta \le 2 \int_{-2\delta}^{2\delta} (1 - \tfrac{|\beta|}{2\delta}) \left| \int_0^N e(u\beta) f(u) du \right|^2 d\beta$$

$$= \delta^{-1} \int_0^N \int_0^N f(u) \overline{f(v)} \left(\frac{\sin 2\pi(u-v)\delta}{\pi(u-v)} \right)^2 du \, dv.$$

Now $|f(u)f(v)| \leq \frac{1}{2}|f(u)|^2 + \frac{1}{2}|f(v)|^2$, so the above is

$$\leq \delta^{-1} \int_0^N \int_0^N |f(u)|^2 \left(\frac{\sin 2\pi(u-v)\delta}{\pi(u-v)}\right)^2 du \, dv$$

$$\leq \delta^{-1} \left(\int_0^N |f(u)|^2 du\right) \left(\int_{-\infty}^{+\infty} \left(\frac{\sin 2\pi v \delta}{\pi v}\right)^2 dv\right)$$

$$= 2 \int_0^N |f(u)|^2 du .$$

We take $f(u) = S(\frac{a}{q}, u)$, $\delta = (qQ)^{-1}$, and use the result (5.7) to obtain

$$\int_{\mathcal{M}_{\frac{a}{q}}} |S(\alpha)|^2 d\alpha \ll (qQ)^{-1} |S(\frac{a}{q})|^2 + (qQ)^{-2} \int_0^N |S(\frac{a}{q}, u)|^2 du .$$

This, with (5.6) gives

$$\rho(1-\rho)NQ^2 \ll Q \sum_{q \leq Q} q^{-1} \sum_{a=1}^{q} {}^* |S(\frac{a}{q})|^2 + \sum_{q \leq Q} q^{-2} \sum_{a=1}^{q} {}^* \int_0^N |S(\frac{a}{q}, u)|^2 du$$

$$\ll Q \sum_{q \leq Q} \left(q \log \frac{2Q}{q}\right)^{-1} \sum_{a=1}^{q} |S(\frac{a}{q})|^2 + \sum_{q \leq Q} q^{-2} \sum_{a=1}^{q} \int_0^N |S(\frac{a}{q}, u)|^2 du$$

$$= Q \sum_{q \leq Q} \left(\log \frac{2Q}{q}\right)^{-1} V_q(N) + \sum_{q \leq Q} q^{-1} \int_0^N V_q(u) \, du$$

$$\leq Q \sum_{q \leq Q} \left(\log \frac{2Q}{q}\right)^{-1} V_q(N) + \sum_{q \leq Q} q^{-1} \sum_{n=1}^{N} V_q(n) .$$

This is (5.4), so the proof is complete.

Corollary 5.2 follows from Theorem 5.1 with $Q = N^{\frac{1}{2}}$, as (5.4) asserts that (5.5) holds on average.

The argument at the bottom of paper 47 and the top of page 48 can be replaced by an appeal to Plancherel's identity. We have

$$\int_{-\delta}^{\delta} \left| \int_0^N e(u\beta) f(u) du \right|^2 d\beta \;\le\; \int_{-\infty}^{+\infty} \left| \int_0^N e(u\beta) f(u) du \right|^2 d\beta$$

$$= \int_0^N |f(u)|^2 du.$$

The argument given can be modified to give a proof of Plancherel's theorem.

Classical mean value theorems

Of the mean value theorems for Dirichlet series, the oldest states that

$$\int_{T_0}^{T_0+T} |S(it)|^2 dt = (T + O(N \log N)) \sum_{n=1}^{N} |a_n|^2, \tag{6.1}$$

where

$$S(s) = \sum_{n=1}^{N} a_n n^{-s}. \tag{6.2}$$

This relation has been used in the theory of the Riemann zeta function (see Titchmarsh, Chapters III, IX).   Drawing on the ideas of the proofs of Lemma 1.5 and Theorem 2.3 we find that we can sharpen (6.1); we have

THEOREM 6.1.   For any real T_0 and T we have

$$\int_{T_0}^{T_0+T} |S(it)|^2 dt = \left(T + \theta \frac{4\pi}{\sqrt{3}} N\right) \sum_{n=1}^{N} |a_n|^2, \tag{6.3}$$

where $S(s)$ is defined by (6.2), and $-1 \le \theta \le 1$.

This result is essentially best possible in the sense that the constant $\frac{4\pi}{\sqrt{3}}$ cannot be replaced by an arbitrarily small number, even when $N = o(T)$.

The q - analogue of the above is the following well-known result.

THEOREM 6.2.   <u>Let</u>

$$S(\chi) = \sum_{n=M+1}^{M+N} a_n \chi(n),\qquad(6.4)$$

<u>where</u> χ <u>is a character modulo</u> q .   <u>Then</u>

$$\sum_{\chi} |S(\chi)|^2 \le \phi(q)\left(1 + \left[\tfrac{N-1}{q}\right]\right) \sum_{\substack{M+1 \\ (n,q)=1}}^{M+N} |a_n|^2.\qquad(6.5)$$

<u>If</u> $N \le q$ <u>then this holds with equality.</u>

Later we make use of

THEOREM 6.3.   <u>Let</u> $S(\chi)$ <u>be defined by</u> (6.4).   <u>If</u> χ <u>is a</u> <u>character modulo</u> q <u>let</u> χ^* <u>denote the primitive character which</u> <u>induces</u> χ .   <u>Then</u>

$$\sum_{\chi} |S(\chi^*)|^2 \le q\left(1 + \left[\tfrac{N-1}{q}\right]\right) \sum_{M+1}^{M+N} |a_n|^2.\qquad(6.6)$$

From the proof it is seen that equality holds when $N = q$ and the a_n are equal.

The methods of proving Theorems 6.1 and 6.2 combine well; we have

THEOREM 6.4.   <u>Let</u>

$$S(s,\chi) = \sum_{n=1}^{N} a_n \chi(n) n^{-s},\qquad(6.7)$$

where χ is a character modulo q.   For any real T_0 and T we have

$$\sum_{\chi} \int_{T_0}^{T_0+T} |S(it,\chi)|^2 dt = \left(\phi(q)T + \theta 4\pi \left(\tfrac{2}{3}\right)^{\frac{1}{2}} N \right) \sum_{\substack{n=1 \\ (n,q)=1}}^{N} |a_n|^2, \qquad (6.8)$$

where $-1 \le \theta \le 1$.

When $N = o(T)$ we obtain a result with a smaller error term if we consider an averaged integral:

THEOREM 6.5   Let $S(s,\chi)$ be defined by (6.7).   For any real T we have

$$\sum_{\chi} \int_{-T}^{T} \left(1 - \tfrac{|t|}{T}\right) |S(it,\chi)|^2 dt = \phi(q) T \left(1 + \theta \, \frac{4\pi^2}{3} \left(\frac{N}{qT}\right)^2 \right) \sum_{\substack{n=1 \\ (n,q)=1}}^{N} |a_n|^2, \quad (6.9)$$

where $-1 \le \theta \le 1$.

The constant $\frac{4\pi^2}{3}$ cannot be replaced by one that is arbitrarily small, but if the a_n are real and non-negative then $\theta \ge 0$ in (6.9), so that $\phi(q) T \sum_{(n,q)=1} |a_n|^2$ is a lower bound for the left hand side, regardless of the size of N.   This observation has some important consequences (see Lemma II.3, and §8.10 of Titchmarsh).

When N is larger than T we can give a result similar to Theorem 6.1, but with the best constant attached to N.   We have

THEOREM 6.6.   Let $S(s)$ be defined by (6.2).   Then for $0 \leq T \leq N$ we have

$$\int_{T_0}^{T_0+T} |S(it)|^2 dt \ = \ 2\pi N \left(1 + O(T^2 N^{-2})\right) \sum_{n=1}^{N} |a_n|^2. \tag{6.10}$$

This bound is asymptotically best possible when

$$a_n = \begin{cases} 0 & 1 \leq n \leq N-h \\ 1 & N-h < n \leq N, \end{cases} \tag{6.11}$$

where $\frac{N}{h} = o(T)$ and $T = o(N)$.   With more work one might expect to prove that

$$\int_{T_0}^{T_0+T} |S(it)|^2 dt \ \leq \ \sum_{n=1}^{N} \left(2\pi n + O(T)\right)|a_n|^2. \tag{6.12}$$

This would be stronger than (6.10), and is asymptotically attained when (6.11) holds and $T = o(N)$, for any $h \leq N$.

As Theorem 6.1 is contained in Theorem 6.4, we begin by proving Theorem 6.2.   If $N \leq q$ then

$$\sum_{\chi} |S(\chi)|^2 \ = \ \phi(q) \sum_{\substack{n \\ (n,q)=1}} |a_n|^2,$$

in view of the identity

$$\sum_{\chi} \chi(a) = \begin{cases} \phi(q) & a \equiv 1 \ (mod\, q) \\ 0 & a \not\equiv 1 \ (mod\, q). \end{cases}$$

If $N > q$ then we break S into $1 + \left[\frac{N-1}{q}\right]$ sums, each of length $\leq q$.   Hence the result.

To prove Theorem 6.3 we note that as in the above proof it suffices to consider $N = q$. The left hand side of (6.6) is

$$= \sum_m \sum_n a_m \overline{a_n} \sum_{\chi} \chi^*(m) \overline{\chi^*}(n).$$

As $|a_m \overline{a_n}| \leq \frac{1}{2}|a_m|^2 + \frac{1}{2}|a_n|^2$, the above is

$$\leq \sum_m |a_m|^2 \sum_n \left| \sum_{\chi} \chi^*(m) \overline{\chi^*}(n) \right|.$$

Let d_1 be the largest divisor of q such that d_1 and (m, q) are composed of precisely the same primes, but not necessarily to the same multiplicities. Let d_2 be determined so that $d_1 d_2$ is the largest divisor of q such that $d_1 d_2$ and (mn, q) are composed of precisely the same primes. Finally, write $d_1 d_2 d_3 = q$. The expression inside the modulus sign above is equal to $\phi(d_3)$ or 0 according as $m \equiv n \pmod{d_3}$ or not. Thus the weight attached to $|a_m|^2$ is

$$= \sum_{\substack{d_2, d_3 \\ d_2 d_3 = \frac{q}{d_1} \\ (d_2, d_3) = 1}} \sum_{\substack{n \\ n \equiv m \pmod{d_3} \\ p | d_2 \Rightarrow p | n}} \phi(d_3).$$

For given pair-wise coprime values of d_1, d_2, d_3 we see that n takes on d_1 values $\bmod\, d_1$, $d_2 \prod_{p | d_2} p^{-1}$ values $\bmod\, d_2$, and 1 value $\bmod\, d_3$. Hence the above is

$$= d_1 \sum_{\substack{d_2, d_3 \\ d_2 d_3 = \frac{q}{d_1} \\ (d_2, d_3) = 1}} \phi(d_3)\, d_2 \prod_{p \mid d_2} p^{-1}$$

$$= q \prod_{p \mid \frac{q}{d_1}} \left(1 - \frac{1}{p} + \frac{1}{p}\right)$$

$$= q .$$

Hence if $q \le N$ then

$$\sum_{\chi} | S(\chi^*)|^2 \le q \sum |a_n|^2,$$

so the result follows.

We now derive Theorems 6.4 and 6.5. We find it simplest to establish Theorem 6.5 first, and then derive Theorem 6.4 from Theorem 6.5. The left hand side of (6.9) is

$$= \phi(q) T \sum_{\substack{n=1 \\ (n, q) = 1}}^{N} |a_n|^2 \; + \; \phi(q) \sum_{\substack{m \ne n \\ m \equiv n (mod\, q) \\ (m, q) = 1}} a_m\, \overline{a_n}\; c(m, n, T), \qquad (6.13)$$

where

$$c(m, n, T) \;=\; T^{-1} \left(\frac{2 \sin \frac{T}{2} \log \frac{n}{m}}{\log \frac{n}{m}}\right)^2 .$$

Now if $1 \le m < n \le N$ then

$$\log \frac{n}{m} \;=\; \log \left(1 - \frac{n-m}{n}\right)^{-1},$$

and by the power series for $\log (1 - x)^{-1}$, this is

$$\geq \frac{n-m}{n}$$

$$\geq \frac{n-m}{N}.$$

Hence

$$0 \leq c(m,n,T) \leq \frac{4N^2}{T(m-n)^2}$$

when $m \neq n$. But $|a_m \overline{a_n}| \leq \frac{1}{2}|a_m|^2 + \frac{1}{2}|a_n|^2$, so the second term in (6.13) is

$$\leq \sum_{\substack{n=1 \\ (n,q)=1}}^{N} |a_n|^2 \frac{8N^2}{Tq^2} \zeta(2)$$

$$= \frac{4\pi^2 N^2}{3Tq^2} \sum_{\substack{n=1 \\ (n,q)=1}}^{N} |a_n|^2.$$

This with (6.13) gives (6.9).

To prove Theorem 6.4 it suffices to show that

$$\sum_{\chi} \int_{-T}^{T} |S(it,\chi)|^2 dt = \frac{\phi(q)}{q}\left(2qT + \theta 4\pi \left(\frac{2}{3}\right)^{\frac{1}{2}} N\right) \sum_{\substack{n=1 \\ (n,q)=1}}^{N} |a_n|^2. \tag{6.14}$$

We let $f(T)$ denote the left hand side of (6.9). Then

$$A^{-1}(T f(T) - (T-A) f(T-A)) \leq \sum_{\chi} \int_{-T}^{T} |S(it,\chi)|^2 dt \leq B^{-1}((T+B) f(T+B) - T f(T)),$$

where $0 \leq A \leq T$ and $0 \leq B$ are chosen below. The above with Theorem 6.5 gives

$$\phi(q)\left(2T - A - \frac{8\pi^2}{3A}\left(\frac{N}{q}\right)^2\right) \leq \sum_{\chi} \int_{-T}^{T} |s(it,\chi)|^2 dt \leq \phi(q)\left(2T + B + \frac{8\pi^2}{3B}\left(\frac{N}{q}\right)^2\right).$$

We take $B = 2\pi\left(\frac{2}{3}\right)^{\frac{1}{2}}\frac{N}{q}$ to obtain the upper bound in (6.14).
If $T < 2\pi\left(\frac{2}{3}\right)^{\frac{1}{2}}\frac{N}{q}$ then the lower bound in (6.14) is negative and
therefore trivial. If $T \geq 2\pi\left(\frac{2}{3}\right)^{\frac{1}{2}}\frac{N}{q}$ then we take $A = 2\pi\left(\frac{2}{3}\right)^{\frac{1}{2}}\frac{N}{q}$,
and the lower bound follows.

Our derivation of Theorem 6.6 is based on equation (1.25).
Here $\nu = \nu(n) = (2\pi)^{-1}\log n$. We take $\delta = (2\pi N)^{-1}$, and
assume that $T \leq N$. The right hand side of (1.25) is
$= 2\pi N \sum |a_n|^2$. For $|t| \leq T$ we see that $|\hat{F}_\delta(t)| = 1 - O(N^2 T^{-2})$,
so the left hand side of (1.25) is

$$\geq \left(1 - O(N^2 T^{-2})\right) \int_{-T}^{T} |s(t)|^2 dt.$$

Hence the result.

CHAPTER 7

New mean value theorems

In the previous chapter we noted that the proofs of the classical Theorems 6.1 and 6.2 combine satisfactorily to give a double average, Theorem 6.4. The proofs of Theorems 2.5 and 6.1, however, do not seem to combine well. Thus we require some new ideas in proving

THEOREM 7.1. <u>Let</u>

$$S(s, \chi) = \sum_{n=1}^{N} a_n \chi(n) n^{-s}, \tag{7.1}$$

<u>where</u> χ <u>is a</u> <u>character</u> <u>modulo</u> q. Then

$$\sum_{q \leq Q} \sum_{\chi}^{*} \int_{T_0}^{T_0+T} |S(it, \chi)|^2 dt \ll (Q^2 T + N) \sum_{n=1}^{N} |a_n|^2 \tag{7.2}$$

<u>for</u> <u>arbitrary</u> $Q, T_0,$ <u>and</u> T.

Previously I obtained [142] a result like this, but weaker by a power of a logarithm. More recently Gallagher (to appear) sharpened the result to its best possible form (7.2). The techniques employed by Gallagher and myself differ quite substantially; we follow Gallagher, as his approach is more elegant and more generally applicable. Gallagher's basic tool is Lemma 1.10; it allows one to combine an integration with another, pre-existing mean value theorem. Gallagher remarks that he was inspired by the work of Fogels [65]. It may be recorded that in addition Titchmarsh [198] (see [205, p. 134-138]) anticipated certain aspects of Gallagher's idea.

Using Gallagher's idea and Theorem 6.3 we obtain

THEOREM 7.2.   Let $S(s,\chi)$ be defined by (7.1).   If χ is a character modulo , let χ^* be the primitive character which induces χ .   Then for arbitrary T_0 and T we have

$$\sum_{\chi} \int_{T_0}^{T_0+T} |S(it, \chi^*)|^2 dt \ll (qT + N) \sum_{n=1}^{N} |a_n|^2. \tag{7.3}$$

Using Gallagher's Lemma 1.10 and Theorem 6.2, we could also give a second proof of the upper bound part of Theorem 6.3, but with larger constants.

We may, of course, use Lemma 1.4 to replace the integral by a finite sum.   We have

THEOREM 7.3.   Let $S(s,\chi)$ be defined by (7.1), and let $\mathcal{T}_\chi$ be a finite set of real numbers in the interval $[T_0 + \frac{\delta}{2}, T_0 + T - \frac{\delta}{2}]$, where $\delta > 0$ is such that for each χ $|t - t'| \geq \delta$ for distinct t and t' in $\mathcal{T}_\chi$.   Then

$$\sum_{q \leq Q} \sum_{\chi}^{*} \sum_{t \in \mathcal{T}_\chi} |S(it, \chi)|^2 \ll (Q^2 T + N)(\delta^{-1} + \log N) \sum_{n=1}^{N} |a_n|^2. \tag{7.4}$$

When $Q = 1$ the above is essentially a result of Davenport (see [142]).   Similarly from Theorem 7.2 we have

THEOREM 7.4.   Under the above hypotheses

$$\sum_{\chi} \sum_{t \in \mathcal{T}_{\chi}} |s(it, \chi)|^2 \ll (qT + N)(\delta^{-1} + \log N) \sum_{n=1}^{N} |a_n|^2. \tag{7.5}$$

For some purposes it is useful to be able to consider a Dirichlet series, not only on a vertical line, but also in a half-plane.   From Theorem 7.3 we deduce the following result, using partial summation.

THEOREM 7.5.   Let $s(s, \chi)$ be as in (7.1), and let $\mathcal{L}_{\chi}$ be a finite set of complex numbers $s = \sigma + it$.   Let T_0, T, σ_0, δ be real numbers such that for each χ

$$T_0 + \frac{\delta}{2} \leqslant t \leqslant T_0 + T - \frac{\delta}{2} \qquad (s \in \mathcal{L}_{\chi}),$$

$$\sigma \geqslant \sigma_0 \qquad (s \in \mathcal{L}_{\chi}),$$

and

$$|t - t'| \geqslant \delta \qquad (s \in \mathcal{L}_{\chi}, s' \in \mathcal{L}_{\chi}, s \neq s').$$

Then

$$\sum_{q \leqslant Q} \sum_{\chi} {}^{*} \sum_{s \in \mathcal{L}_{\chi}} |s(s, \chi)|^2 \ll (Q^2 T + N)(\delta^{-1} + \log N) \cdot$$

$$\cdot \sum_{n=1}^{N} |a_n|^2 n^{-2\sigma_0} \left(1 + \log \frac{\log 2N}{\log 2n}\right). \tag{7.6}$$

Similarly from Theorem 7.4 we have

THEOREM 7.6. <u>Under the above hypotheses</u>

$$\sum_{\chi} \sum_{s \in d_{\chi}} |S(s,\chi)|^2 \ll (qT+N)(\delta^{-1}+\log N) \cdot$$

$$\cdot \sum_{n=1}^{N} |a_n|^2 \, n^{-2\sigma_0} \left(1+\log \frac{\log 2N}{\log 2n}\right). \qquad (7.7)$$

We now give Gallagher's proof of Theorem 7.1.   If $0 \leqslant T \leqslant 1$ then we apply Theorem 2.5 pointwise to obtain the result.   If $T \geqslant 1$ then by Lemma 1.10

$$\sum_{q \leqslant Q} \sum_{\chi} \,^{*}\int_{T_0}^{T_0+T} |S(it,\chi)|^2 dt \ll T^2 \sum_{q \leqslant Q} \sum_{\chi} \,^{*}\int_{0}^{\infty} |\sum_{y}^{\tau y} a_n \chi(n) \, n^{-iT_0}|^2 \frac{dy}{y},$$

and by Theorem 2.5 this is

$$\ll T^2 \int_{0}^{\infty} (Q^2+(\tau-1)y)\left(\sum_{y}^{\tau y} |a_n|^2\right) \frac{dy}{y}.$$

Now $|a_n|^2$ receives a weight which is $\ll Q^2 T + n$, so Theorem 7.1 follows.   The proof of Theorem 7.2 is the same, except that it is based on Theorem 6.3 instead of Theorem 2.5.

To derive Theorem 7.3 we note that from Lemma 1.4

$$\sum_{t \in \mathcal{T}_{\chi}} |S(it,\chi)|^2 \leq \delta^{-1}\int_{T_0}^{T_0+T} |S(it,\chi)|^2 dt + \left(\int_{T_0}^{T_0+T} |S(it,\chi)|^2 dt\right)^{\frac{1}{2}}\left(\int_{T_0}^{T_0+T} |S'(it,\chi)|^2 dt\right)^{\frac{1}{2}}.$$

Hence by Cauchy's inequality

$$\sum_{q \leq Q} \sum_{\chi}^{*} \sum_{t \in \mathcal{T}_\chi} |S(it,\chi)|^2 \ll \delta^{-1} \sum_{q \leq Q} \sum_{\chi}^{*} \int_{T_0}^{T_0+T} |S(it,\chi)|^2 dt$$

$$+ \left(\sum_q \sum_{\chi}^{*} \int |S(it,\chi)|^2 dt \right)^{\frac{1}{2}} \left(\sum_q \sum_{\chi}^{*} \int |S'(it,\chi)|^2 dt \right)^{\frac{1}{2}},$$

and by Theorem 7.1 this is

$$\ll (Q^2 T + N)(\delta^{-1} + \log N) \sum_{n=1}^{N} |a_n|^2,$$

so (7.3) follows.   Theorem 7.4 follows similarly.

To prove Theorem 7.5 we let

$$S(s,\chi,u) = \sum_{2 \leq n \leq u} a_n \chi(n) n^{-s},$$

so that

$$S(s,\chi) = a_1 + S(\sigma_0+it,\chi,N) N^{\sigma_0-\sigma} + (\sigma-\sigma_0) \int_2^N S(\sigma_0+it,\chi,u) u^{-\sigma+\sigma_0-1} du.$$

Hence by Cauchy's inequality

$$|S(s,\chi)|^2 \ll |a_1|^2 + |S(\sigma_0+it,\chi)|^2 + (\sigma-\sigma_0)^2 \left(\int_2^N (\log u) u^{-2\sigma+2\sigma_0-1} du \right) \cdot$$

$$\cdot \left(\int_2^N |S(\sigma_0+it,\chi,u)|^2 u^{-1} (\log u)^{-1} du \right)$$

$$\ll |a_1|^2 + |S(\sigma_0+it,\chi)|^2 + \int_2^N |S(\sigma_0+it,\chi,u)|^2 (u \log u)^{-1} du.$$

This with (7.4) gives

$$\sum_{\gamma \leq Q} \sideset{}{^*}\sum_{x} \sum_{s \in d_x} |S(s,x)|^2 \ll (Q^2 T + N)(\delta^{-1} + \log N) \cdot$$

$$\cdot \left(|a_1|^2 + \sum_{n=2}^{N} |a_n|^2 n^{-2\sigma_0} \left(1 + \int_{n}^{N} (u \log u)^{-1} du\right)\right)$$

$$\ll (Q^2 T + N)(\delta^{-1} + \log N) \sum_{n=1}^{N} |a_n|^2 n^{-2\sigma_0} \left(1 + \log \frac{\log 2N}{\log 2n}\right),$$

so the proof is complete.   The proof of Theorem 7.6 is the same, except that we use (7.5) instead of (7.4).

Large moduli theorems

Our mean value theorems can be used to give upper bounds for the number of times a Dirichlet series is large. The corresponding upper bound for a Fourier series is essentially best possible, but for a Dirichlet series much more is true. Halász [73] (see also [74],[75]) established this for certain special Dirichlet series, although his method applies (see [142]) to an arbitrary Dirichlet polynomial. Gallagher suggested that Halász's method might be abstracted to a Hilbert space; Lemma 1.7 is the appropriate abstraction. Subsequently Bombieri found Lemma 1.5, which is more satisfactory from most points of view. Lemma 1.5 gives us immediately

LEMMA 8.1.   Let

$$S(s,\chi) = \sum_{n=1}^{N} a_n \chi(n) n^{-s},\tag{8.1}$$

where χ is a character modulo q, and let $\mathcal{L}$ be a finite set of triples (s,q,χ). Then

$$\sum_{(s,q,\chi)\in\mathcal{L}} |S(s,\chi)|^2 \ll \left(\sum_{n=1}^{N} |a_n|^2 b_n^{-1}\right) \max_{(s,q,\chi)\in\mathcal{L}} \sum_{(s',q',\chi')\in\mathcal{L}} |B(\bar{s}+s', \bar{\chi}\chi')|,\tag{8.2}$$

where

$$B(s,\chi) = \sum_{n=1}^{\infty} b_n \chi(n) n^{-s}\tag{8.3}$$

is absolutely convergent at the points $\bar{s}+s'$, and the b_n

are non-negative real numbers for which $b_n > 0$ whenever $a_n \neq 0$.

An appropriate choice of the b_n gives

THEOREM 8.2.   Let $S(s,\chi)$ be given by (8.1), and let $\mathcal{L}$ be a finite set of triples (s, q, χ) , where χ is primitive. Let σ_0 be such that

$$\sigma \geq \sigma_0 \tag{8.4}$$

for each triple $(\sigma + it, q, \chi) \in \mathcal{L}$.   Let δ be such that

$$|t - t'| \geq \delta \tag{8.5}$$

for any distinct triples $(\sigma + it, q, \chi)$ and $(\sigma' + it', q, \chi)$ in $\mathcal{L}$.   Let $T \geq 2$ be such that

$$|t| \leq T \tag{8.6}$$

for any triple $(\sigma + it, q, \chi)$ in $\mathcal{L}$.   Then

$$\sum_{(s,q,\chi) \in \mathcal{L}} |S(s,\chi)|^2 \ll (1 + \delta^{-1})(N + |\mathcal{L}| Q T^{\frac{1}{2}} \log QT) \sum_{n=1}^{N} |a_n|^2 n^{-2\sigma_0}. \tag{8.7}$$

THEOREM 8.3.   Let $S(s,\chi)$ be given by (8.1), and let $\mathcal{L}$ be a finite set of triples (s, d, χ) where χ is primitive modulo d and $d | q$.   Let σ_0 and $T \geq 2$ satisfy (8.4) and (8.6) for any $(s, d, \chi) \in \mathcal{L}$, and let δ satisfy (8.5) for any distinct (s, d, χ) , (s', d, χ) in $\mathcal{L}$.   Then

$$\sum_{(s,d,\chi) \in \mathcal{L}} |S(s,\chi)|^2 \ll (1 + \delta^{-1})(N + |\mathcal{L}| q^{\frac{1}{2}} T^{\frac{1}{2}} \log q T) \sum_{n=1}^{N} |a_n|^2 n^{-2\sigma_0}. \tag{8.8}$$

If we use alternative estimates for $|B(s,\chi)|$ we obtain various modifications of (8.7). If we take $Q = 1$ then we can obtain estimates for $|B(s)|$ in terms of

$$M(\alpha, T) = \max_{\substack{\sigma \geq \alpha \\ |t| \leq T \\ |s-1| \geq 1}} |\gamma(s)|. \tag{8.9}$$

We have

THEOREM 8.4. <u>Let</u>

$$S(s) = \sum_{n=N}^{2N} a_n n^{-s}, \tag{8.10}$$

<u>and</u> <u>suppose</u> <u>that</u> $\mathcal{A}$ <u>is a</u> <u>finite</u> <u>set</u> <u>of</u> <u>complex</u> <u>numbers</u> $s = \sigma + it$ <u>such</u> <u>that</u> (8.4) <u>and</u> (8.6) <u>hold</u> <u>for</u> <u>all</u> $s \in \mathcal{A}$, <u>and</u> <u>such</u> <u>that</u> (8.5) <u>holds</u> <u>for</u> <u>distinct</u> s <u>and</u> s' <u>in</u> $\mathcal{A}$. <u>Then</u>

$$\sum_{s \in \mathcal{A}} |S(s)|^2 \ll (1+\delta^{-1})(N+|\mathcal{A}|M(\theta, 4T)N^\theta) \sum_{n=N}^{2N} |a_n|^2 n^{-2\sigma}, \tag{8.11}$$

<u>where</u> $0 \leq \sigma \leq 1$.

If $|S(s,\chi)| \geq V$ for all $(s, \gamma, \chi) \in \mathcal{A}$ then (8.7) tells us that

$$|\mathcal{A}| \ll (1+\delta^{-1}) NV^{-2} \sum_{n=1}^{N} |a_n|^2 n^{-2\sigma}, \tag{8.12}$$

provided that

$$V^2 \geq A(1+\delta^{-1}) QT^{t}(\log QT) \sum_{n=1}^{N} |a_n|^2 n^{-2\sigma}, \tag{8.13}$$

where A is a sufficiently large absolute constant. The

inequality (8.11) also gives us (8.12), but now the condition
is that

$$V^2 \geqslant A(1+\delta^{-1}) M(\theta, 4T) N^\theta \sum_{n=N}^{2N} |a_n|^2 n^{-2\sigma_0}, \qquad (8.14)$$

where A is a sufficiently large absolute constant.   We see
that (8.12) is a much stronger bound than what we would obtain
from a mean value theorem.   On the other hand the conditions
(8.13) and (8.14) are stronger than we would like.   We
conjecture that (8.12) holds provided only that

$$V^2 \geqslant (1+\delta^{-1})(NQT)^\varepsilon \sum_{n=1}^{N} |a_n|^2 n^{-2\sigma_0}. \qquad (8.15)$$

It seems clear that (8.12) is within a constant of being best
possible.

If we were interested in characters to a fixed modulus
then we could show (by modifying the proof of Theorem 8.2)
that

$$\sum_{(s,\chi) \in \mathscr{S}} |S(s,\chi)|^2 \ll (1+\delta^{-1})(N+|\Delta|^{\frac{1}{2}} T^{\frac{t}{2}} \log_g T) \sum_{n=1}^{N} |a_n|^2 n^{-2\sigma_0}. \qquad (8.16)$$

It seems likely that some generalization to algebraic number
fields can be made.   In this situation $B(\bar{s}+s', \bar{\chi}\chi')$ becomes
a "scalor product" in the sense of A.I. Vinogradov [219] and
Moroz [145], [146].

We now prove Lemma 8.1.   In Lemma 1.5 we take
$\xi = \{a_n b_n^{-\frac{1}{2}}\}$, $\varphi = \{b_n^{\frac{1}{2}} \bar{\chi}(n) n^{-\sigma+it}\}$.   Then
$\|\xi\|^2 = \sum |a_n|^2 b_n^{-1}$ and $(\varphi, \varphi') = B(\bar{s}+s', \bar{\chi}\chi')$, so the

result follows.

To prove Theorem 8.2 we take $b_n = \left(1 - \frac{n}{2N}\right) n^{2\sigma_0}$ for $1 \le n \le 2N$, and $b_n = 0$ for $n > 2N$. In Appendix II we show that

$$|B(s,x)| \ll \left(q(|t|+2)\right)^{\frac{1}{2}} \log q(|t|+2) + \varepsilon(x) N(|t|+2)^{-2}, \qquad (8.17)$$

when $\sigma \ge 0$. Here $B(s,x)$ is given by (8.3), and $\varepsilon(x)$ is 1 or 0 according as x is principal or non-principal. In Theorem 8.2 the characters are primitive, so $\bar{x}x'$ is principal if and only if $x = x'$. When $x = x'$ we have (8.5), so the terms $N(|t|+2)^{-2}$ contribute an expression of the sort

$$N \sum_i (|t_i|+2)^{-2} \ll N(1+\delta^{-1}) \int_{-\infty}^{+\infty} (|u|+2)^{-2} du \ll N(1+\delta^{-1}).$$

Hence Theorem 8.2 follows from Lemma 8.1   We derive Theorem 8.3 in the same way; here $\bar{x}x'$ is a character modulo the least common multiple of d and d'.

To deduce Theorem 8.4 we take $b_n = \left(e^{-\frac{n}{2N}} - e^{-\frac{n}{N}}\right) n^{2\sigma_0}$ in Lemma 8.1. In Appendix II we show that

$$|B(s)| \ll M(\theta, 2T) N^\theta + N e^{-|t|} \qquad (8.18)$$

when $\sigma \ge 0$. Here $B(s)$ is given by (8.3) with $x(n) = 1$ for all n, and $0 \le \theta \le 1$. Now Theorem 8.4 follows from Lemma 8.1.

Further results and conjectures
concerning mean and large moduli

In Chapters 2 and 4 we have already mentioned ways in
which the large sieve might be sharpened.   With $S(\mathbf{x})$ defined
by (2.2), Burgess [29] has found that

$$\sum_{\mathbf{q} \in \mathcal{2}} \sum_{a=1}^{\mathbf{q}} {}^* |S(\tfrac{a}{\mathbf{q}})| \ll (Q X)^{\frac{1}{2}} (N + Q X)^{\frac{1}{2}} \left(\sum_{M+1}^{M+N} |a_n|^2 \right)^{\frac{1}{2}}, \qquad (9.1)$$

where $\mathcal{2}$ is a set of X numbers, each of which is $\leq Q$.   This
result has the disadvantage that it treats only the first
power of $S(\mathbf{x})$, but it has the advantage that the bound is
sharp when N is as large as $Q X$, a considerable gain if $\mathcal{2}$
is sparse, so that QX is considerably smaller than Q^{2}.   The
bound (9.1) may be derived, using Cauchy's inequality, from
Theorem 2.1 with $\delta = (Q X)^{\frac{1}{2}}$ and the observation that

$$\sum_{\mathbf{q} \in \mathcal{2}} \sum_{a=1}^{\mathbf{q}} {}^* N_{\frac{1}{\mathbf{q} X}} (\tfrac{a}{\mathbf{q}}) \ll Q X.$$

Burgess and Elliott (to appear) have used (9.1) with $\mathcal{2}$ a set
of squares, to give a bound for the average of the least
primitive root modulo the square of a prime.

Elliott [56] has recently found mean value theorems like
Theorem 2.5, but in which the average is restricted to
characters of a given order.   Typical among his results is the
inequality

$$\sum_{p \leqslant x} \left| \sum_{n \leqslant H} a_n \left(\tfrac{n}{p} \right) \right|^2 \ll x \sum_{\substack{mn = t^2 \text{ or } 2t^2 \\ m \leqslant H \\ n \leqslant H}} |a_m a_n| + H \log H \left(\sum_{n \leqslant H} |a_n| \right)^2. \qquad (9.2)$$

This can be expected to be sharp only when H is smaller than $x^{\frac{1}{2}}$; it would be nice to have a result like this which is sharp when H is as large as x. We could prove a result of the desired sort if we could show that (8.12) holds under the assumption (8.15). Elliott [51], [52], [53], [56] has also found a number of interesting applications of his mean value theorems. Lavrik [120] has also given results of the same general nature as Elliott's.

The most important shortcoming in the previous chapters is the absence of a proof that (8.15) implies (8.12). What we seem to require (aside from the Lindelöf hypothesis) is a better bound for the bilinear form (1.17) which occurs in the proof of Lemma 1.5. The treatment we gave was essentially a proof of a theorem of Browne [23], and with some modification it provides a proof of a theorem of Farnell [62]. These theorems are related to others of A. Brauer [22], Ostrowski [147], Parker [149], and Perron [150]; see also [138]. There are cases in which these theorems give best possible results, but it seems that they are inferior for our present purposes. That the loss in our argument does not occur elsewhere is perhaps not clear from the treatment we have given, but Bombieri has recently observed that in the present case a bound for the bilinear form (1.17) is _equivalent_ to having the same bound for the left hand side of (1.12). In fact Lemma 1.5 is closely related to the following well-known result:

Let $C = \{c_{mn}\}$ be an $M \times N$ matrix such that

$$\sum_{n=1}^{N} \left| \sum_{m=1}^{M} c_{mn} x_m \right|^2 \leq K \sum_{m=1}^{M} |x_m|^2 \tag{9.3}$$

for any values of the x_m. Then also

$$\sum_{m=1}^{M} \left| \sum_{n=1}^{N} c_{mn} y_n \right|^2 \leq K \sum_{n=1}^{N} |y_n|^2 \tag{9.4}$$

for any values of the y_n.

The above statement is obviously self-dual, so that a bound for the left hand side of (9.3) is equivalent to a bound for the left hand side of (9.4). In more generality, if A is a bounded linear operator on a Banach space B and if A^* is the adjoint operator on the dual space B^* then $\|A\| = \|A^*\|$. In the particular case at hand we require a bound for

$$\sum_{s \in \mathcal{A}} \left| \sum_{n} a_n n^{-s} \right|^2, \tag{9.5}$$

and without any loss we may give instead a bound for

$$\sum_{n} \left| \sum_{s \in \mathcal{A}} c_s n^{-s} \right|^2. \tag{9.6}$$

Here $\mathcal{A}$ is a finite set of complex numbers with $\sigma \geq 0, |t| \leq T$ for all $s \in \mathcal{A}$, and $|t - t'| \geq 1$ for distinct s and s' in $\mathcal{A}$.

Squaring out, we see that the above is

$$= \sum_{s \in \mathcal{A}} \sum_{s' \in \mathcal{A}} c_s \overline{c_{s'}} \sum_{n=1}^{N} n^{-s-\overline{s'}} \tag{9.7}$$

In other words we require a bound for an $|\mathcal{A}| \times |\mathcal{A}|$ bilinear form in which the diagonal entries are N and the non-diagonal entries may be assumed to be (on the Lindelöf hypothesis) no larger than $N^{\frac{1}{2}} T^{\varepsilon}$. Now a <u>random</u> $R \times R$ matrix with unimodular entries is bounded by $R^{\frac{1}{2}+\varepsilon}$, so we may expect that the form (9.7) is

$$\ll (N + |\mathcal{A}|^{\frac{1}{2}+\varepsilon} N^{\frac{1}{2}} T^{\varepsilon} + |\mathcal{A}|) \sum_{s \in \mathcal{A}} |c_s|^2$$

$$\ll (N + |\mathcal{A}|^{1+\varepsilon} T^{\varepsilon}) \sum_{s \in \mathcal{A}} |c_s|^2.$$

Hence we form

CONJECTURE 9.1. <u>If</u> $\mathcal{A}$ <u>is a finite set of complex numbers</u> <u>which fits the above description, then</u>

$$\sum_{s \in \mathcal{A}} \left| \sum_{n=1}^{N} a_n n^{-s} \right|^2 \ll (N + |\mathcal{A}|^{1+\varepsilon} T^{\varepsilon}) \sum_{n=1}^{N} |a_n|^2, \tag{9.8}$$

<u>for any</u> N <u>and any values of the</u> a_n.

A consequence of this conjecture is that (8.15) would imply (8.12). A q-analogue of this conjecture may also be formulated, and also a conjecture combining characters and n^{-s}. The above conjecture, if true, would seem to be one of our deepest statements concerning the behavior of arbitrary

Dirichlet series. A weaker conjecture, one which is perhaps easier to prove, we state as

CONJECTURE 9.2. Let

$$S(s) = \sum_{n=1}^{N} a_n n^{-s},$$

and let the numbers $a_n(k)$ be defined by the relation

$$\left(S(s)\right)^k = \sum_{n=1}^{N^k} a_n(k) n^{-s}.$$

Then

$$\int_0^T |S(it)|^{2\nu} dt \ll (T+N^\nu) \left(\sum_{n=1}^{N^2} |a_n(2)|^2 \right)^{\frac{\nu}{2}} \tag{9.9}$$

uniformly for $1 \leqslant \nu \leqslant 2$.

In applications it would suffice to have (9.9) with the right hand side increased by a factor N^ε. In this weaker form Conjecture 9.2 is an easy consequence of Conjecture 9.1. We note that Theorem 6.1 implies that

$$\int_0^T |S(it)|^{2k} dt = (T+O(N^k)) \sum_{n=1}^{N^k} |a_n(k)|^2,$$

so (9.9) holds when $\nu = 1$ or $\nu = 2$. In applications it would be useful to know that (9.9) holds also when $1 < \nu < 2$, especially when $N^\nu \approx T$.

C H A P T E R   10

Mean moduli of L-functions

In 1918 Hardy and Littlewood [80] proved that

$$\int_0^T |\zeta(\tfrac{1}{2}+it)|^2 dt \sim T \log T.$$

Subsequently Littlewood[135], Ingham [96], Titchmarsh[201], and Atkinson [3] replaced the right hand side above by more explicit expressions. In 1926 Ingham [96] showed that

$$\int_0^T |\zeta(\tfrac{1}{2}+it)|^4 dt \sim \frac{1}{2\pi^2} T (\log T)^4.$$

Titchmarsh[198] immediately gave a second proof of this, using ideas that anticipate Gallagher's Lemma 1.9. Subsequently Kober[103], Titchmarsh[202], Atkinson [2], and Bellman [10] elaborated on Titchmarsh's approach. In 1931 Paley [148] proved that

$$\sum_{\chi} |L(\tfrac{1}{2}+it,\chi)|^4 \ll_{t,\varepsilon} q^{1+\varepsilon}.$$

Haselgrove [83] proved a similar result for an integral of the fourth power of an L-function, and Linnik [130](see also [114]) showed that

$$\sum_{\chi} |L(\tfrac{1}{2}+it,\chi)|^4 \ll \phi(q)(|t|+2)(\log q(|t|+2))^9.$$

Later Gallagher [68] gave a simple proof that

$$\sum_{q \leq Q} \sum_{\chi}^{*} |L(\tfrac{1}{2}+it,\chi)|^4 \ll Q^2 (|t|+2)^2 \left(\log q \, (|t|+2)\right)^4.$$

Huxley [93] has combined aspects of Linnik's and Gallagher's results in giving an average upper bound for sixth and eighth powers of L-functions. To these results we add

THEOREM 10.1. If $T \geq 2$ then

$$\sum_{\chi}^{*} \int_{-T}^{T} |L(\tfrac{1}{2}+it,\chi)|^4 dt \ll \phi(q) T (\log q T)^4 \tag{10.1}$$

when

$$\tfrac{1}{2} - (\log q T)^{-1} \leq \sigma \leq \tfrac{1}{2} + (\log q T)^{-1}. \tag{10.2}$$

From this we deduce

COROLLARY 10.2. If $T \geq 2$ then

$$\sum_{\chi}^{*} \int_{-T}^{T} |L(\tfrac{1}{2}+it,\chi) L'(\tfrac{1}{2}+it,\chi)|^2 dt \ll \phi(q) T (\log q T)^6. \tag{10.3}$$

These results and Lemma 1.4 enable us to establish

THEOREM 10.3. For each primitive character $\chi \bmod q$ let $\mathcal{T}_\chi$ be a finite set of real numbers lying in the interval $[-T, T], T \geq 2$, and let δ be so small that

$$|t - t'| \geq \delta$$

for any χ and any distinct t and t' in $\mathcal{T}_\chi$. Then

$$\sum_{\chi}^{*} \sum_{t \in \mathcal{T}_{\chi}} |L(\tfrac{1}{2}+it, \chi)|^{4} \ll (\delta^{-1} + \log q T) \phi(q) T (\log q T)^{4}. \qquad (10.4)$$

A result of this sort has already appeared [143], but the proof found there is unsatisfactory.   The fault lies in the use of the approximate functional equation of Chandrasekharan and Narasimhan [33], in which the implied constants are not absolute.   The corresponding result of Lavrik [119] does not have this disadvantage;   we correct the error by appealing to his work instead of that of Chandrasekharan and Narasimhan.

By summing (10.4) over divisors of q we obtain

COROLLARY 10.4.   Let the above hypotheses hold.   If χ is a character mod q let χ^{*} denote the primitive character which induces χ.   Then

$$\sum_{\chi} \sum_{t \in \mathcal{T}_{\chi}} |L(\tfrac{1}{2}+it, \chi^{*})|^{4} \ll (\delta^{-1} + \log q T) q T (\log q T)^{4}. \qquad (10.5)$$

The powers of logarithms in (10.1), (10.3), and (10.4) are all best possible.   In fact our upper bounds could probably be replaced by asymptotic relations in all but our last result.   In deriving these results we use the mean value theorems of Chapter 6, but for these results to be applicable we must first show that $L(s, \chi)^{2}$ can be approximated by Dirichlet polynomials.

Writers requiring formulae for L-functions have often
had to derive their own results, but recent work provides
formulae for almost any conceivable purpose.   Chandrasekharan
and Narasimhan [33] have shown that in principle a wide class
of functions possess approximate functional equations.
However, the bounds they obtain for their error terms are in
many cases larger than one would expect the main terms to be,
so their results are to some extent only of formal interest.
On the other hand Lavrik [115], [116], [117], [118], [119] and
Huxley [93] have proceeded in the same spirit, and have
obtained very precise results.   We use one of Lavrik's results
to obtain

LEMMA 10.5.   Let χ be a primitive character $mod\, q$,
and let $s = \sigma + it$ satisfy

$$\tfrac{1}{2} - \frac{2}{\log q\tau} \leq \sigma \leq \tfrac{1}{2} + \frac{2}{\log q\tau}, \tag{10.6}$$

where $\tau = \max(1, |s|)$. Then

$$|L(s,\chi)|^4 \ll (q\tau)^{-1} \int_{\frac{q\tau}{8\pi}}^{\frac{2q\tau}{\pi}} \Big| \sum_{n \leq u} a(x,u)\,\chi(n)\,n^{-s} \Big|^2 du$$

$$+ (q\tau)^{-1} \int_{\frac{q\tau}{8\pi}}^{\frac{2q\tau}{\pi}} \Big| \sum_{n \leq v} a(n,v)\,\overline{\chi}(n)\,n^{s-1} \Big|^2 dv \tag{10.7}$$

$$+ R_1(s,\chi),$$

where $|a(n,u)| \leq d(n)$ and $R_1(s,\chi) \geq 0$ is such that

$$\sum_{\chi}^{*} R_1(s,\chi) \ll \phi(q)(\log q\,\tau)^3. \tag{10.8}$$

We now prove this lemma. From Theorem 1 of Lavrik[119] we see that if s and q are as above and if

$$\frac{1}{2}\left(\frac{q\tau}{2\pi}\right)^{\frac{1}{2}} \le x \le 2\left(\frac{q\tau}{2\pi}\right)^{\frac{1}{2}}, \quad xy = \frac{q\tau}{2\pi}, \tag{10.9}$$

then

$$|L(s,\chi)|^4 \ll \left|\sum_{n\le x}\chi(n)n^{-s}\right|^4 + \left|\sum_{n\le y}\overline{\chi}(n)n^{s-1}\right|^4 + R_2(s,\chi,x), \tag{10.10}$$

where

$$R_2(s,\chi,x) = 1 + \left|\sum_{n\le(q\tau)^{\frac{1}{2}}\log q\tau} c(n;s,\chi(-1),x)\chi(n)\right|^4$$

$$+ \left|\sum_{n\le(q\tau)^{\frac{1}{2}}\log q\tau} c(n;1-s,\chi(-1),y)\overline{\chi}(n)\right|^4. \tag{10.11}$$

Here $c(n;s,\pm1,x)$ is a function for which

$$|c(n;s,\pm1,x)| \ll (q\tau)^{-\frac{1}{4}}\left(1 + \tau^{\frac{1}{2}}\left|1-\left(\frac{2\pi nx}{q\tau}\right)^2\right|\right)^{-1}\exp\left(\frac{-n^2}{2q\tau}\right) \tag{10.12}$$

uniformly for σ and x satisfying (10.6) and (10.9). We write

$$\left(\sum_{n\le x}\chi(n)n^{-s}\right)^2 = \sum_{n\le x^2}a(n,x^2)\chi(n)n^{-s},$$

put u in place of x^2, and integrate both sides of (10.10) with respect to u. This gives us (10.7), where $R_1(s,\chi)$ is the average of $R_2(s, \chi)$.

We now must show that (10.8) holds. Here the main difficulty lies in showing that

$$\sum_{\chi} R_2(s, \chi, x) \ll \phi(q) q^{-1} \tau^{-\frac{1}{2}} \sum_{n \le Aq\tau} d(n)^2 \left(1 + \tau^{\frac{3}{2}} \left|1 - n^{t} \left(\frac{2\pi x}{q\tau}\right)\right|^3\right)^{-1}$$

$$+ \phi(q) \left(\log q\tau\right)^3$$

(10.13)

for x in the range (10.9). The bound (10.8) follows easily from (10.13), for if we replace x^2 by u in (10.13) and average we find that

$$R_1(s, \chi) \ll \phi(q) q^{-1} \tau^{-1} \sum_{n \le Aq\tau} d(n)^2 \int_{-\infty}^{+\infty} \left(1 + |v|^3\right)^{-1} dv$$

$$+ \phi(q) \left(\log q\tau\right)^3$$

$$\ll \phi(q) \left(\log q\tau\right)^3.$$

This is (10.8), so to complete the proof of the lemma it suffices to prove (10.13).

We first consider

$$\sum_{\chi} \left| \sum_{n \le (q\tau)^{\frac{1}{2}} \log q\tau} c(n; s, 1, x) \chi(n) \right|^4.$$

This expression is

$$\ll \sum_{i=1,2,3} \sum_{\chi} \left| \sum_{n\,(i)} c(n;3,1,x)\, \chi(n) \right|^4, \tag{10.14}$$

where the $\sum_{n}(i)$ denote sums over the ranges

$$1 \le n \le \frac{q^{\tau}}{4\pi x}\,, \quad \frac{q^{\tau}}{4\pi x} < n \le \frac{q^{\tau}}{\pi x}\,, \quad \frac{q^{\tau}}{\pi x} < n \le (q\tau)^{\frac{1}{2}} \log q\tau,$$

respectively. From Theorem 6.2 and (10.12) we have

$$\sum_{\chi} \left| \sum_{n\,(1)} \right|^4 \ll \phi(q)(q+q\tau)q^{-2}\,\tau^{-3} \sum_{n \le \left(\frac{q\tau}{4\pi x}\right)^2} d(n)^2$$

$$\ll \phi(q)\,\tau^{-1}(\log q\tau)^3$$

$$\ll \phi(q)(\log q\tau)^3. \tag{10.15}$$

To treat $\sum_{n}(3)$ we divide the sum into subsums $\sum_{(3,j)}$ over the ranges $\frac{q\tau 2^{j-1}}{\pi x} < n \le \frac{q\tau 2^j}{\pi x}$. From Hölder's inequality

$$\left| \sum_{n\,(3)} \right|^4 \ll \sum_{j} 2^{3j} \left| \sum_{(3,j)} \right|^4. \tag{10.16}$$

From Theorem 6.2 and (10.12) we see that if $U \ge \frac{q\tau}{\pi x}$ then

$$\sum_{\chi} \left| \sum_{n=U}^{2U} c(n;3,1,x)\,\chi(n) \right|^4 \ll \phi(q)(q+U^2)q^{-2}\,\tau^{-3}\,U^2(\log U)^3$$
$$\cdot \exp\left(-\frac{U^2}{2q\tau}\right).$$

Hence from (10.16) we have

$$\sum_{\chi} \left| \sum_{n}(3) \right|^4 \ll \phi(q)(\log q\tau)^3. \tag{10.17}$$

To treat $\left| \sum_{n}(2) \right|^4$ we divide the sum into $\ll \log \tau$ subsums $\sum_{n}(2,j)$ in which n runs over the intervals $I_j = (a_j, b_j]$, as follows:

$$I_j = \left(\frac{q\tau}{2\pi x}(1+2^{j-1}\tau^{-\frac{1}{2}}), \frac{q\tau}{2\pi x}(1+2^{j}\tau^{-\frac{1}{2}}) \right] \qquad (j > 0),$$

$$I_0 = \left(\frac{q\tau}{2\pi x}(1 - \tau^{-\frac{1}{2}}), \frac{q\tau}{2\pi x}(1 + \tau^{-\frac{1}{2}}) \right],$$

$$I_{-j} = \left(\frac{q\tau}{2\pi x}(1 - 2^{j}\tau^{-\frac{1}{2}}), \frac{q\tau}{2\pi x}(1 - 2^{j-1}\tau^{-\frac{1}{2}}) \right] \qquad (j > 0).$$

Again by Hölder's inequality

$$\left| \sum_{n}(2) \right|^4 \ll \sum_{j} 2^{|j|} \left| \sum_{n}(2,j) c(n;s,1,x) \chi(n) \right|^4.$$

For n in I_j we have $c(n;s,1,x) \ll (q\tau)^{-\frac{1}{4}} 2^{-|j|}$, so by Theorem 6.2 we have

$$\sum_{\chi} \left| \sum_{n}(2,j) \right|^4 \ll \phi(q)(q + q\tau^{\frac{1}{2}}2^{|j|})q^{-2}\tau^{-1}2^{-4|j|} \sum_{a_j^2}^{b_j^2} d(n)^2$$

$$\ll \phi(q)q^{-1}\tau^{-\frac{1}{2}}2^{-4|j|} \sum_{a_j^2}^{b_j^2} d(n)^2.$$

Hence

$$\sum_{\chi} \left| \sum_{n} (2) \right|^4 \ll \phi(q) q^{-1} \tau^{-\frac{1}{2}} \sum_{j} 2^{-3|j|} \sum_{a_j^2}^{b_j^2} d(n)^2$$

$$\ll \phi(q) q^{-1} \tau^{-\frac{1}{2}} \sum_{n} d(n)^2 \left(1 + \tau^{\frac{3}{2}} \left| n^{\frac{1}{2}} \left(\frac{2\pi x}{j \tau} \right) - 1 \right|^3 \right)^{-1},$$

so from (10.14), (10.15), (10.17) and the above we have

$$\sum_{\chi}^{*} \left| \sum_{n} c(n; s, 1, x) \chi(n) \right|^4 \ll \phi(q) q^{-1} \tau^{-1} \sum_{n} d(n)^2 \left(1 + \tau^{\frac{3}{2}} \left| n^{\frac{1}{2}} \left(\frac{2\pi x}{j \tau} \right) - 1 \right|^3 \right)^{-1}$$

$$+ \phi(q) (\log q \tau)^3.$$

The same treatment applies to $c(n; s, -1, x)$, so (10.13) holds, and (10.8) follows.

The above proof is based on Lavrik's [119] proof of his Corollary 3, but our treatment is somewhat simplified. In particular we have avoided making an appeal to a result of Linnik and A.I. Vinogradov [132] concerning the size of $\sum d(n)^2$ summed over short intervals. An averaged approximate functional equation would enable one to make further simplifications: the estimate corresponding to (10.12) would be much sharper, so that our treatment of $\left| \sum_{n} (2) \right|^4$ would be rendered redundant.

Recalling that

$$\sum_{n \leq x} d(n)^2 \, n^{-1} \ll (\log x)^4 ,$$

we see that Theorem 10.1 follows immediately from Lemma 10.5 and Theorem 6.3.

To deduce Corollary 10.2 we use the identity

$$f'(a) = \frac{1}{2\pi i} \int f(z)(z-a)^{-2} dz$$

from which by the Cauchy - Schwarz inequality

$$|f'(a)|^2 \ll r^{-3} \int |f(z)|^2 |dz| . \tag{10.18}$$

Here the integration is around a circle of radius r and center a . We take $r = (\log_9 T)^{-1}$ and $f(s) = L(s, \chi)^2$ in (10.18), and find that

$$\int_{-T}^{T} |L(\tfrac{1}{2}+it, \chi) L'(\tfrac{1}{2}+it, \chi)|^2 dt \ll (\log_9 T)^3 \int_{\tfrac{1}{2}-(\log_9 T)^{-1}}^{\tfrac{1}{2}+(\log_9 T)^{-1}} \int_{-T-1}^{T+1} |L(\sigma+it, \chi)|^4 dt \, d\sigma .$$

Hence (10.3) follows from Theorem 10.1.

We use Lemma 1.4 to derive Theorem 10.3.   We have

$$\sum_{\chi}^{*} \sum_{t \in \mathcal{J}_{\chi}} |L(\tfrac{1}{2}+it, \chi)|^4 \ll \sum_{\chi}^{*} \delta^{-1} \int_{-2T}^{2T} |L(\tfrac{1}{2}+it, \chi)|^4 dt$$

$$+ \sum_{\chi}^{*} \left(\int_{-2T}^{2T} |L(\tfrac{1}{2}+it, \chi)|^4 dt \right)^{\tfrac{1}{2}} \left(\int_{-2T}^{2T} |L(\tfrac{1}{2}+it, \chi) L'(\tfrac{1}{2}+it, \chi)|^2 dt \right)^{\tfrac{1}{2}} .$$

By Cauchy's inequality the last term above is

$$\leq \left(\sum_{x}^{*} \int_{-2T}^{2T} |L(\tfrac{1}{2}+it,x)|^4 dt \right)^{\frac{1}{2}} \left(\sum_{x}^{*} \int_{-2T}^{2T} |L(\tfrac{1}{2}+it,x)L'(\tfrac{1}{2}+it,x)|^2 dt \right)^{\frac{1}{2}}.$$

Now (10.4) follows from (10.1), (10.3), and the above.

C H A P T E R   11

Zero-free regions and the proliferation of zeros

In this chapter we consider the possibility of the
existence of a zero $\rho = \beta + i\gamma$ of $\zeta(s)$ for which β is near 1 ,
where by "near" we mean $1 - \beta \leq A(\log\log |\gamma|)^{-1}$, with A a
certain absolute constant. We find that if β is near 1 then
there are other zeros nearby. A quantitative formulation
provides a new method of deducing zero-free regions for $\zeta(s)$.
Our results can be extended to L-functions, but for simplicity
we restrict ourselves to the zeta function. A simple result
of the sort we have in mind is

THEOREM 11.1. $\zeta(\rho_0) = 0, \rho_0 = \beta_0 + i\gamma_0 , \gamma_0 > 0,$ _and_
if

$$1 - \beta_0 \leq (\log\log \gamma_0)^{-1}, \tag{11.1}$$

then there are

$$\gg (1 - \beta_0)^{-1}(\log\log \gamma_0)^{-1} \tag{11.2}$$

zeros $\rho = \beta + i\gamma$ of $\zeta(s)$ in the rectangle $\frac{3}{4} \leq \beta \leq 1$,
$\frac{\gamma_0}{3} \leq \gamma \leq 3\gamma_0$.

Recently Levinson [12] showed that if ρ_0 is a an isolated
zero of $\zeta(s)$ then (11.1) does not hold; his result is similar to
the above, and is contained in Therem 11.2. On the other hand it
is not clear that his technique can be used to give a good lower
bound, such as (11.2), for the number of nearby zeros.

Our method allows us to give a more precise localization of the zeros induced by ρ_0. Let $n(t, w, h)$ denote the number of zeros $\rho = \beta + i\gamma$ of $\mathfrak{I}(s)$ in the rectangle $1 - w \leq \beta \leq 1$, $t - \frac{h}{2} < \gamma \leq t + \frac{h}{2}$. We would like to be able to assert that if (11.1) holds then

$$n(\gamma_0, \delta, \delta) \gg \delta(1 - \beta_0)^{-1}$$

uniformly for $1 - \beta_0 \leq \delta \leq (\log\log\gamma_0)^{-1}$. We fail to obtain this in two respects. Firstly, our use of the relation $3 + 4\cos\theta + \cos 2\theta \geq 0$ introduces the possibility that some or all of the induced zeros are in the vicinity of $1 + 2i\gamma_0$ instead of $1 + i\gamma_0$. Secondly, we are able to show only that a suitable average of $n(\gamma_0, w, h) + n(2\gamma_0, w, h)$ is as large as we expect. Our basic result is

THEOREM 11.2.   If $\mathfrak{I}(\rho_0) = 0$, $\beta_0 > \frac{1}{2}$, $\gamma_0 > 0$, then for $1 - \beta_0 \leq \delta \leq 1$

$$\delta^2 \int_0^1 \int_0^\infty (n(\gamma_0, w, h) + n(2\gamma_0, w, h))(h+\delta)^{-5} \exp(-w\delta^{-1}) \, dh \, dw \gg (1 - \beta_0)^{-1}. \qquad (11.3)$$

From (11.2) and (11.3) we see that the number of zeros induced by ρ_0 is at least proportional to $(1 - \beta_0)^{-1}$. In the opposite direction, an upper bound for $|\mathfrak{I}(s)|$ and Jensen's formula can be used to obtain an upper bound for $n(t, w, h)$, and hence an upper bound for $(1 - \beta_0)^{-1}$. We recall that

$$\mathfrak{I}(s) \ll (1 + T^{\frac{1}{2}(1-\sigma)}) \log T, \qquad (11.4)$$

where $0 \leq \sigma \leq 2$, and $T = |t| + 2$. From this and Theorem 11.2

we deduce

COROLLARY 11.3 (de la Vallée-Poussin). There is an absolute constant $c > 0$ such that if $\Im(\rho) = 0$, $\rho = \beta + i\gamma$, $\tau = |\gamma| + 2$, then

$$\beta \leq 1 - c(\log \tau)^{-1}. \tag{11.5}$$

Richert [162], following the work of I.M. Vinogradov and others, has shown that

$$\Im(s) \ll \left(1 + T^{100(1-\sigma)^{\frac{3}{2}}}\right)(\log T)^{\frac{2}{3}}, \tag{11.6}$$

where $0 \leq \sigma \leq 2$, $T = |t| + 2$. From this we deduce

COROLLARY 11.4. There is an absolute constant $c > 0$ such that if $\Im(\rho) = 0$, $\rho = \beta + i\gamma$, $\tau = |\gamma| + 2$, then

$$\beta \leq 1 - c(\log \tau)^{-\frac{2}{3}}(\log\log \tau)^{-\frac{1}{3}}. \tag{11.7}$$

The above (see [226, p. 226]) has for several years been the best zero-free region for the zeta-function. Previous proofs of (11.5) have used either the "global" results of Hadamard (see §3.8 of Titchmarsh or section 13 of Davenport), or the "local" method of Landau [107] (see §3.9 of Titchmarsh). Our proof of (11.7) does not depend on Landau's function-theoretic lemmas, unlike the previous proofs of sharpenings of (11.5).

Before proving our basic Theorem 11.2, we show that Theorem

11.1 follows from Theorem 11.2.   In (11.3) we take
$\delta = \frac{1}{5}(\log\log \gamma_0)^{-1}$.   We may suppose that $1-\beta_0 \leq \delta$, for if
$\frac{1}{5}(\log\log \gamma_0)^{-1} \leq 1-\beta_0 \leq (\log\log \gamma_0)^{-1}$ then (11.2) holds
trivially.   A zero $\rho = \beta + i\gamma$ for which $\gamma < \frac{1}{3}\gamma_0$ or $\gamma > 3\gamma_0$
contributes an amount $\ll (\gamma - \gamma_0)^{-4}$ to the left hand side of
(11.3).   In view of the fact (see Titchmarsh, Theorem 9.2)
that

$$n(t, 1, 1) \ll \log(|t|+1), \tag{11.8}$$

the totality of such zeros contribute an amount $\ll \gamma_0^{-2}$, which
is negligible.   A zero $\rho = \beta + i\gamma$ for which $\beta < \frac{3}{4}$ contributes
at most

$$\ll (\log \gamma_0)^{-\frac{5}{4}} \min\left(\delta^{-2}, \ \delta^2\left((\gamma-\gamma_0)^{-4} + (\gamma - 2\gamma_0)^{-4}\right)\right)$$

so again from (11.8) we see that such zeros contribute

$$\ll (\log \gamma_0)^{-\frac{1}{4}} (\log\log \gamma_0)^2$$

$$\ll 1$$

to the left hand side of (11.3).   Thus the zeros in the
rectangle $\frac{3}{4} \leq \beta \leq 1$, $\frac{1}{3}\gamma_0 \leq \gamma \leq 3\gamma_0$ contribute the major portion
of the left hand side of (11.3).   But no zero contributes more
than $O(\delta^{-1})$, so there must be $\gg \delta(1-\beta_0)^{-1} \gg (1-\beta_0)^{-1}(\log\log \gamma_0)^{-1}$
of them; this is (11.2).

We now prove Theorem 11.2.   For $\sigma > 1$ we write

$$f(s; x, \delta) = \mathrm{Re} \sum_{n > x} \Lambda(n) n^{-s} \left(1 - \left(\frac{x}{n}\right)^\delta\right)^3, \tag{11.9}$$

so that

$$3 f(\sigma; x, \delta) + 4 f(\sigma + it; x, \delta) + f(\sigma + 2it; x, \delta)$$

$$= 2 \sum_{n > x} \Lambda(n) n^{-\sigma} \left(1 - \left(\tfrac{x}{n}\right)^{\delta}\right)^{3} (1 + \cos(t \log n))^{2}$$

$$\geqslant 0 \qquad\qquad (11.10)$$

for $\sigma > 1$. Now $f(s; x, \delta)$ depends on the zeros of the zeta function; we state the precise result as

LEMMA 11.5. For $\sigma > 1, x > 1$ we have

$$f(s; x, \delta) = -6 \delta^{3} \mathcal{R}e \sum_{\rho} x^{\rho - s} (s - \rho)^{-1} (s + \delta - \rho)^{-1} (s + 2\delta - \rho)^{-1} (s + 3\delta - \rho)^{-1}$$

$$+ 6 \delta^{3} \mathcal{R}e \; x^{1 - s} (s - 1)^{-1} (s + \delta - 1)^{-1} (s + 2\delta - 1)^{-1} (s + 3\delta - 1)^{-1}$$

$$+ O(\delta^{3} x^{-2}). \qquad\qquad (11.11)$$

Proof. For any s and any $x > 1$ we have

$$\sum_{n \leq x}' \Lambda(n) n^{-s} = -\frac{\zeta'}{\zeta}(s) + \sum_{\rho} x^{\rho - s} (s - \rho)^{-1}$$

$$- x^{1 - s} (s - 1)^{-1} + \sum_{n=1}^{\infty} x^{-2n - s} (s + 2n)^{-1}, \qquad (11.12)$$

with the obvious interpretation if there are infinite terms. This formula is well-known; the case $s = 0$ is the usual explicit formula for $\Psi(x)$, and the more general result is obtained by the same method. For $\sigma > 1$

$$-\frac{\zeta'}{\zeta}(s) = \sum_{n=1}^{\infty} \Lambda(n) n^{-s},$$

so from (11.12) we have

$$\sum_{n>x}' \Lambda(n)\, n^{-s} = - \sum_{\rho} x^{s-\rho}(s-\rho)^{-1} + x^{s-1}(s-1)^{-1}$$

$$- \sum_{n=1}^{\infty} x^{-2n-s}(2n+s)^{-1}$$

for $\sigma > 1$. We take this at the points s, $s+\delta$, $s+2\delta$, and $s+3\delta$. This gives (11.11), in view of the identity

$$\sum_{\nu=0}^{3} (-1)^{\nu}\binom{3}{\nu}(z+\nu\delta)^{-1} = 6\,\delta^3(z(z+\delta)(z+2\delta)(z+3\delta))^{-1}.$$

As a consequence of Lemma 11.5 we have the inequality

$$f(\sigma;x,\delta) \le (\sigma-1)^{-1} + O(\delta^3) \tag{11.13}$$

for $x > 1$, $\sigma > 1$. To estimate the contribution of the zeros in (11.11) we have

LEMMA 11.6. If $\operatorname{Re} z \ge 0$ then

$$\operatorname{Re} 6\delta^3 \exp(-z\delta^{-1})\,(z(z+\delta)(z+2\delta)(z+3\delta))^{-1} \ge -\frac{9}{\delta}. \tag{11.14}$$

Proof: The left hand side is

$$= \operatorname{Re} \exp(-z\delta^{-1})\,z^{-1} - 3\operatorname{Re} \exp(-z\delta^{-1})(z+\delta)^{-1} + 3\operatorname{Re} \exp(-z\delta^{-1})(z+2\delta)^{-1}$$

$$- \operatorname{Re} \exp(-z\delta^{-1})(z+3\delta)^{-1}.$$

Now $|z+\nu\delta|^{-1} \le (\nu\delta)^{-1}$ when $\operatorname{Re} z \ge 0$, so the last three terms contribute an amount that is $\ge -5\delta^{-1}$. If $|z| \ge \delta$ then the first term is $\le \delta^{-1}$ in modulus, and we are done. If $|z| \le \delta$

then we write

$$z^{-1} \exp\left(-z\delta^{-1}\right) = z^{-1}\left(\exp\left(-z\delta^{-1}\right) - 1\right) + z^{-1}.$$

The real part of the second term is non-negative, and by the maximum modulus principle the modulus of the first term on the right is $\leq 4\delta^{-1}$ in the disc $|z| \leq \delta$. The lemma now follows.

Using our two lemmas we now complete the proof of Theorem 11.2. From (11.10), (11.11), and (11.13) we have

$$8\delta^3 \operatorname{Re} \sum_{\rho \neq \rho_0} x^{\rho-s}\left((s-\rho)(s+\delta-\rho)(s+2\delta-\rho)(s+3\delta-\rho)\right)^{-1}$$

$$+ 2\delta^3 \operatorname{Re} \sum_{\rho} x^{\rho-s}\left((\sigma+2it-\rho)(\sigma+2it+\delta-\rho)(\sigma+2it+2\delta-\rho)(\sigma+2it+3\delta-\rho)\right)^{-1}$$

$$\leq (\sigma-1)^{-1} - 8\delta^3 \operatorname{Re} x^{\rho_0-s}\left((s-\rho_0)(s+\delta-\rho_0)(s+2\delta-\rho_0)(s+3\delta-\rho_0)\right)^{-1}$$

$$+ O(\delta^3). \tag{11.15}$$

We write $\rho_0 = \beta_0 + i\gamma_0$, and suppose that β_0 is near 1 and that γ_0 is large and positive. We take $\sigma = 1 + 18(1-\beta_0)$, $t = \gamma_0$, and $x = \exp\left(-\delta^{-1}\right)$. Then

$$(\sigma-\beta_0)^{-1} = \frac{18}{19}(\sigma-1)^{-1}, \tag{11.16}$$

and for $\delta \geq 40.19.(1-\beta_0)$ we have

$$x^{\beta_0-\sigma} = \exp\left(-19\delta^{-1}(1-\beta_0)\right)$$

$$\geq \exp\left(-(40)^{-1}\right)$$

$$\geq \frac{19}{20}. \tag{11.17}$$

Also

$$6\delta^3\left((\sigma+\delta-\beta_0)(\sigma+2\delta-\beta_0)(\sigma+3\delta-\beta_0)\right)^{-1} = \left(1+19(1-\beta_0)\delta^{-1}\right)^{-1}\left(1+\tfrac{19}{2}(1-\beta_0)\delta^{-1}\right)^{-1}\left(1+\tfrac{19}{3}(1-\beta_0)\delta^{-1}\right)^{-1}$$

$$\geqslant \left(1+\tfrac{1}{40}\right)^{-1}\left(1+\tfrac{1}{80}\right)^{-1}\left(1+\tfrac{1}{120}\right)^{-1}$$

$$\geqslant \frac{20}{21}. \tag{11.18}$$

Multiplying together the respective sides of (11.16), (11.17), and (11.18), we see that

$$8\delta^3\, Re\; x^{\beta_0-s}\left((s-\rho_0)(s+\delta-\rho_0)(s+2\delta-\rho_0)(s+3\delta-\rho_0)\right)^{-1} \geqslant \frac{8}{7}(\sigma-1)^{-1},$$

so the right hand side of (11.15) is

$$\leqslant -\frac{1}{7}(\sigma-1)^{-1} + O(\delta^3) \leqslant \left(200(1-\beta_0)\right)^{-1}$$

if $\delta \leqslant 7$ and $1-\beta_0$ is sufficiently small.

We now consider the left hand side of (11.15). If ρ is within δ of $1+i\gamma_0$ or $1+2i\gamma_0$ then we apply Lemma 11.6 to the term. We estimate the contribution of the other zeros trivially. Altogether we now have

$$\delta^3 \sum_{|\rho-1-i\gamma_0|>\delta} x^{\rho-1}\left(|\gamma-\gamma_0|+\delta\right)^{-4} + \delta^3 \sum_{|\rho-1-2i\gamma_0|>\delta} x^{\rho-1}\left(|\gamma-2\gamma_0|+\delta\right)^{-4}$$

$$+ \sum_{|\rho-1-i\gamma_0|\leqslant\delta} \delta^{-1} + \sum_{|\rho-1-2i\gamma_0|\leqslant\delta} \delta^{-1} \gg (1-\beta_0)^{-1}.$$

The left hand side of this is majorized by the left hand side of (11.3), so the proof is complete.

To deduce the Corollaries we suppose that

$$\delta \le (\log |\gamma_0|)^{-\frac{1}{4}}.$$ (11.19)

In view of (11.8) the zeros at a distance more than 1 from $1 + i\gamma_0$ or $1 + 2i\gamma_0$ contribute a negligible amount to the left hand side of (11.3). Hence from (11.3) we see that if (11.19) holds then there exists an r, $\delta \le r \le 1$, for which

$$n(\gamma_0, r, r) + n(2\gamma_0, r, r) \gg \frac{r^3}{\delta^2(1-\beta_0)}.$$ (11.20)

On the other hand we recall that Jensen's inequality asserts that if f is an analytic function whose modulus is bounded by M in the disc $|z| \le r$, then the number of zeros of f in $|z| \le \frac{r}{2}$ is no more than $2 \log (M|f(0)|^{-1})$. Hence from (11.4) and (11.6) we have the upper bounds

$$n(t, r, r) \ll r \log t + \log\log t$$ (11.21)

and

$$n(t, r, r) \ll r^{\frac{3}{2}} \log t + \log\log t$$ (11.22)

for $t \ge 4$.

To obtain Corollary 11.3 we take $\delta = (\log \gamma_0)^{-1} \log\log \gamma_0$. This satisfied (11.19), and for $r \ge \delta$ the second term on the right hand side of (11.21) may be ignored. From (11.20) and (11.21) we have

$$(1-\beta_0)^{-1} \ll \delta^2 r^{-2} \log \gamma_0$$

$$\ll \log \gamma_0,$$

which gives (11.5).

To obtain Corollary 11.4 we take $\delta = (\log \gamma_0)^{-\frac{3}{2}} (\log\log \gamma_0)^{\frac{2}{3}}$
This satisfies (11.19), and for $r \geq \delta$ the second term on the
right hand side of (11.22) may be ignored.   From (11.20) and
(11.22) we have

$$(1-\beta_0)^{-1} \ll \delta^2 r^{-\frac{3}{2}} \log \gamma_0$$

$$\ll \delta^{\frac{1}{2}} \log \gamma_0$$

$$\ll (\log \gamma_0)^{\frac{2}{3}} (\log\log \gamma_0)^{\frac{1}{3}},$$

which gives (11.7).

C H A P T E R   12

Distribution of zeros of L-functions

We now consider bounds for the number $N(\sigma, T, \chi)$ of zeros $\rho = \beta + i\gamma$ of the function $L(s, \chi)$ in the rectangle $\sigma \leq \beta \leq 1$, $-T \leq \gamma \leq T$. If χ is principal then we write $N(\sigma, T, \chi) = N(\sigma, T)$ for the number of zeros of the zeta function in this rectangle.

In 1914 Bohr and Landau [12] (see also Littlewood [136]) proved that for $\sigma > \frac{1}{2}$

$$N(\sigma, T) \ll_\sigma T, \tag{12.1}$$

and as it was already known that $N(\frac{1}{2}, T) \gg T \log T$, the above established that almost all zeros of $\zeta(s)$ are near the line $\sigma = \frac{1}{2}$. In retrospect we see that this much is true of certain other functions, for example, the functions

$$\zeta_Q(s) = \sum_{(x,y) \neq (0,0)} Q(x,y)^{-s},$$

where $Q(x,y) = ax^2 + bxy + cy^2$, $b^2 - 4ac < 0$, a, b, c integers. These functions possess functional equations, but in general they have zeros to the right of the line $\sigma = \frac{1}{2}$ (see [153]). In fact Davenport and Heilbronn [49] proved (see also Cassels [32]) that if Q is a positive definite quadratic form of discriminant $d < 0$ and if the class number $h(d)$ exceeds 1, then the Epstein zeta function $\zeta_Q(s)$ has a zero in the half plane $\sigma > 1$. From this it follows that there is a $\delta = \delta(Q) > 0$ for which

$$N_Q (1+\delta, T) \gg_Q T$$

for $T > T_o(Q)$.   Hence a bound of the sort (12.1) is best
possible for these functions.   Using the Euler product for
$\zeta(s)$, Bohr and Landau [13] improved (12.1) to

$$N(\sigma, T) = o_\sigma (T).$$

Subsequently Carlson [31] (see also Hoheisel [87] and
Landau [106]) showed that

$$N(\sigma, T) \ll T^{4\sigma(1-\sigma)+\varepsilon} \qquad\qquad (\tfrac{1}{2} \le \sigma \le 1),$$

Hoheisel [88] that

$$N(\sigma, T) \ll T^{4\sigma(1-\sigma)}(\log T)^6 \qquad\qquad (\tfrac{1}{2}+\delta \le \sigma \le 1),$$

Titchmarsh [199] that

$$N(\sigma, T) \ll T^{\frac{4(1-\sigma)}{3-2\sigma}+\varepsilon}, \qquad\qquad\qquad (12.2)$$

and Ingham [97] that

$$N(\sigma, T) \ll T^{2(1+2c)(1-\sigma)}(\log T)^5, \qquad\qquad (12.3)$$

provided that $\zeta(\tfrac{1}{2}+it) \ll t^c \log t$ for $t \ge 2$.   Later Ingham
[98] added to this the inequality

$$N(\sigma, T) \ll T^{\frac{3(1-\sigma)}{2-\sigma}}(\log T)^5. \qquad\qquad (12.4)$$

Selberg [187], [191] sharpened these results when σ is near
$\tfrac{1}{2}$, and Turán [207], [209], [210], [211], [212], [213] has
proved a number of results, some of them hypothetical.   In

particular he has proved [2/2] that

$$N(\sigma, T) \ll T^{2(1-\sigma)+(1-\sigma)^{1.14}} \tag{12.5}$$

when $1 - \delta \leq \sigma \leq 1$. Iglina [95] has given similar estimates when σ is near 1 . More recently Halász and Turán [75] used Richert's bound (11.6) to show that for σ near 1

$$N(\sigma, T) \ll T^{(1-\sigma)^{\frac{3}{2}}(\log \frac{1}{1-\sigma})^3} . \tag{12.6}$$

Here the essential idea was provided by Halász; this idea is embodied in the inequality (1.14). Bombieri [17] has unified and extended the results of Halász and Turán.

Many results corresponding to the above have been established for the zeros of L-functions, though the situation is complicated by the fact that various bounds have been given for $N(\sigma, T, \chi)$, $\sum_{\chi} N(\sigma, T, \chi)$, and $\sum_{q \leq Q} \sum_{\chi}^{*} N(\sigma, T, \chi)$. Linnik [127], [128], Čudakov [43] , [44] , and Haselgrove [93] have each given bounds for $N(\sigma, T, \chi)$, though their bounds are only slightly stronger than known bounds for $\sum_{\chi} N(\sigma, T, \chi)$. A bound of the sort

$$\sum_{\chi} N(\sigma, T, \chi) \ll q^{c(1-\sigma)} \qquad (\tfrac{1}{2} \leq \sigma \leq 1), \tag{12.7}$$

for some range of T , is essential to the proof of Linnik's theorem on the least prime in an arithmetic progression. Results of this sort have been given by Linnik [123], [124],

[125] , Rodosskiĭ [171], [172] , Turán [214] , Fogels [65] , and Gallagher [71] . The important feature of (12.7) is that it is strong when σ is near 1 . Selberg [190] proved a result which is correspondingly strong when σ is near $\frac{1}{2}$. Turán [208], Rodosskiĭ [167], [168], [169],[170], Tatuzawa [197], and Pospeev [152] have given bounds of the general form

$$\sum_{\chi} N(\sigma, T, \chi) \ll q^{\phi(\sigma)} T^{\Theta(\sigma)} (\log q T)^A . \tag{12.8}$$

We prove

THEOREM 12.1. Suppose that $q \geq 1$ and $T \geq 2$. For $\frac{1}{2} \leq \sigma \leq \frac{4}{5}$ we have

$$\sum_{\chi} N(\sigma, T, \chi) \ll (q T)^{\frac{3(1-\sigma)}{2-\sigma}} (\log q T)^9 , \tag{12.9}$$

and for $\frac{4}{5} \leq \sigma \leq 1$ we have

$$\sum_{\chi} N(\sigma, T, \chi) \ll (q T)^{\frac{2(1-\sigma)}{\sigma}} (\log q T)^{14} . \tag{12.10}$$

These bounds strengthen all of the previous bounds of the sort (12.8). We note that (12.9) is a natural extension of (12.4). The bound (12.10) is new even when $q = 1$.

In 1948 Rényi [155] showed that there is a constant $\delta > 0$ such that no non-principal L-function mod p has a zero $\rho = \beta + i\gamma$ in the rectangle $1 - \delta \leq \beta \leq 1$, $|\gamma| \leq (\log p)^3$, for all but $\ll A^{\frac{3}{4}}$ primes in the range $A \leq p \leq 2A$. It was recognised (see Barban [4]) that a sharpening of this would be useful; in 1965 A.I. Vinogradov [220] and Bombieri [15] independently gave strengthenings of Rényi's bound. Vinogradov proved (essentially) that

$$\sum_{q \leq Q} \sum_{\chi}{}^{*} N(\sigma, T, \chi) \ll Q^{3-2\sigma+\varepsilon} (T \log Q)^{C\varepsilon^{-4}}, \qquad (12.11)$$

while Bombieri showed that

$$\sum_{q \leq Q} \sum_{\chi}{}^{*} N(\sigma, T, \chi) \ll T(Q^2 + QT)^{\frac{4(1-\sigma)}{3-2\sigma}} (\log QT)^{10}. \qquad (12.12)$$

The bounds (12.11) and (12.12) may be used to prove an important theorem concerning the distribution of prime numbers in arithmetic progressions (see Chapter 15). Elliott [50] has found an additional use for (12.12). Bombieri used his form of the large sieve in proving (12.12); Vinogradov's proof of (12.11) is not so easily described. Using Theorems 7.5, 8.2, and 10.3, we obtain

THEOREM 12.2. _Suppose that_ $Q \geq 1$ _and_ $T \geq 2$. _For_ $\frac{1}{2} \leq \sigma \leq \frac{4}{5}$ _we have_

$$\sum_{q \leq Q} \sum_{\chi}{}^{*} N(\sigma, T, \chi) \ll (Q^2 T)^{\frac{3(1-\sigma)}{2-\sigma}} (\log QT)^{9}, \qquad (12.13)$$

while for $\frac{4}{5} \leq \sigma \leq 1$ _we have_

$$\sum_{q \leq Q} \sum_{\chi}{}^{*} N(\sigma, T, \chi) \ll (Q^2 T)^{\frac{2(1-\sigma)}{\sigma}} (\log QT)^{14}. \qquad (12.14)$$

An account of this result has already appeared [143];
simultaneously Jutila [101] gave similar but weaker results,
with interesting applications.

By using information concerning the size of $L(\frac{1}{2}+it, \chi)$,
such as Burgess's bounds [26], [28], one can obtain better
results when σ is near 1 (see Iglina [95]).   For simplicity
we treat in detail only $\zeta(s)$, about which we prove

THEOREM 12.3.   Let

$$M(\alpha, T) = \max_{\substack{Re\, s \geq \alpha \\ |Im\, s| \leq T \\ |s-1| \geq 1}} |\zeta(s)|. \tag{12.15}$$

Then for $T \geq 2$

$$N(\sigma, T) \ll \left(A M(\alpha, 8T)(\log T)^5 \right)^{\frac{2(1-\sigma)(3\sigma-1-2\alpha)}{(2\sigma-1-\alpha)(\sigma-\alpha)}} (\log T)^8, \tag{12.16}$$

provided that $\frac{1}{2} \leq \alpha \leq 1$ and $\sigma \geq \frac{1+\alpha}{2}$.   Here A is a large
absolute constant.

The case $\alpha = \frac{1}{2}$ has already been published [143].
Because of arithmetic consequences, an inequality of the sort

$$N(\sigma, T) \ll T^{2(1-\sigma)}(\log T)^A \tag{12.17}$$

or sometimes

$$N(\sigma, T) \ll T^{2(1-\sigma)+\varepsilon} \tag{12.18}$$

has been labelled the "density hypothesis".   The following

result contains the density hypothesis in the restricted range $\frac{9}{10} \leqslant \sigma \leqslant 1$.

COROLLARY 12.4. For $\sigma > \frac{3}{4}$ and $T > 2$ we have

$$N(\sigma, T) \ll T^{\frac{4(1-\sigma)(3\sigma-2)}{3(4\sigma-3)(2\sigma-1)}} (\log T)^{50}. \qquad (12.19)$$

Ingham's inequality (12.3) shows that the density hypothesis is a consequence of the Lindelöf hypothesis, which asserts that $M(\frac{1}{2}, T) < T^{\varepsilon}$. More recently Turán [213] demonstrated that the density hypothesis would follow from a certain Hypothetical Lemma. This Hypothetical Lemma is a consequence of the Lindelöf hypothesis, but may be easier to establish. Bombieri [17] modified the proof of Theorem 12.3 to show that if Turán's Hypothetical Lemma is true then (12.16) may be replaced by

$$N(\sigma, T) \ll_{\alpha} (M(2T))^{\frac{2(1-\sigma)}{2\sigma-1-\alpha}} T^{\varepsilon}$$

for $\sigma > \frac{1+\alpha}{2}$. Taking $\alpha = -\varepsilon^{-1}$ gives Turán's result.

Some time ago Turán wrote that one ought to be able to deduce the estimate

$$N(\sigma, T) < T^{\varepsilon} \qquad\qquad (\varepsilon = \varepsilon(\sigma), \sigma > \tfrac{1}{2}) \quad (12.20)$$

from the Lindelöf hypothesis. Subsequently Halász and Turán [75] showed that this is so for $\sigma > \frac{3}{4}$. Their Theorem 1 states that if $|\zeta(\frac{1}{2}+it)| < t^{\varepsilon}$ for $t > t_0(\varepsilon)$ then

$$N\left(\tfrac{3}{4} + 2\varepsilon^{\frac{1}{5}}, t\right) < t^{3\varepsilon^{\frac{1}{2}}}.$$

We see that Theorem 12.3 with $\alpha = \tfrac{1}{2}$ is sharper. From Theorem 12.3 we have also

COROLLARY 12.5.   If $\sigma \geqslant \tfrac{1}{2}$, $T \geqslant 2$ then

$$N(\sigma, T) \ll T^{167(1-\sigma)^{\frac{3}{2}}} (\log T)^{17}. \tag{12.21}$$

This bound is sharper than (12.6), except in a region contained in the zero-free region given in Corollary 11.4.

Conjectures 9.1 and 9.2 can be brought to bear on the problem of estimating $N(\sigma, T)$.   An examination of our proof of Theorem 12.3 makes it clear that if Conjecture 9.1 is valid then

$$N(\sigma, T) \ll \left(M(2T) \, T^{\frac{2}{2\sigma-1}} \right)^{\frac{4(1-\sigma)}{2\sigma-1}} (\log T)^{A} \tag{12.22}$$

for $\sigma > \tfrac{1}{2}$.   Hence Conjecture 9.1 and the Lindelöf hypothesis together imply that (12.20) holds.   In addition we prove

THEOREM 12.6.   If Conjecture 9.2 is valid then

$$N(\sigma, T) \ll T^{2(1-\sigma)} (\log T)^{16} \tag{12.23}$$

for $\tfrac{1}{2} \leqslant \sigma \leqslant 1$, $T \geqslant 2$.

Thus Conjecture 9.2, which seems so weak, can be used to

derive the density hypothesis in the strong form (12.17).

We now establish the machinery necessary for the proofs of Theorems 12.1 and 12.2. In proving results of this sort it has been customary to appeal to a function-theoretic lemma of Littlewood (see Davenport, §25) and to a mean value theorem of Gabriel [67]. This approach could be adopted in proving (12.9) and (12.13), but not (12.10) and (12.14). The method we develop is simple, and direct, and enables us to give a unified treatment.

We first observe that in Theorems 12.1 and 12.2 we may assume that qT and Q^2T , respectively, are larger than prescribed absolute constants. To simplify the writing of logarithms we assume that the parameters X and Y , introduced below, satisfy $2 \leqslant X \leqslant Y \leqslant (qT)^A$ or $2 \leqslant X \leqslant Y \leqslant (Q^2T)^A$, respectively, where A is an absolute constant. When we assign values to X and Y it is easy to check that these inequalities hold.

As is usual in these problems, we let

$$M_\chi(s,\chi) = \sum_{n \leqslant X} \mu(n)\, \chi(n)\, n^{-s}.$$

The Dirichlet series for $L(s,\chi)M_\chi(s,\chi)$ has coefficients $a_n\, \chi(n)$, where

$$a_n = a_n(\chi) = \sum_{\substack{d \mid n \\ d \leqslant X}} \mu(d).$$

Thus $a_1 = 1$, $a_n = 0$ for $2 \leqslant n \leqslant X$, and $|a_n| \leqslant d(n)$ for

$n > X$.   Using a well-known Mellin transform, we have

$$e^{-\frac{1}{Y}} + \sum_{n>X} a_n \chi(n) n^{-s} e^{-\frac{n}{Y}} = \sum_{n=1}^{\infty} a_n \chi(n) n^{-s} e^{-\frac{n}{Y}}$$

$$= \frac{1}{2\pi i} \int_{c-i\infty}^{c+i\infty} L(s+w, \chi) M_\chi(s, \chi) Y^w \Gamma(w) dw.$$

We now suppose that $\frac{1}{2} < \sigma < 1$.   We take the contour to the line $\mathcal{R}e\, w = \frac{1}{2} - \sigma$, and in doing so we pass a simple pole at $w = 0$, and also one at $w = 1-s$ if χ is a principal character.   Our equation becomes

$$e^{-\frac{1}{Y}} + \sum_{n>X} a_n \chi(n) n^{-s} e^{-\frac{n}{Y}} = L(s, \chi) M_\chi(s, \chi) \tag{12.24}$$

$$+ \varepsilon(\chi) \frac{\phi(q)}{q} M_\chi(1, \chi) Y^{1-s} \Gamma(1-s)$$

$$+ \frac{1}{2\pi} \int_{-\infty}^{+\infty} L(\tfrac{1}{2}+it+iu, \chi) M_\chi(\tfrac{1}{2}+it+iu, \chi) \cdot$$

$$\cdot Y^{\frac{1}{2}-\sigma+iu} \Gamma(\tfrac{1}{2}-\sigma+iu) du,$$

where $\varepsilon(\chi) = 1$ when χ is principal and $\varepsilon(\chi) = 0$ otherwise.
We may expect that if X and Y are moderately large then
$e^{-\frac{1}{Y}} \doteq 1 \doteq L(s, \chi) M_\chi(s, \chi)$ and that the other three terms are
small.   This is in fact the case if the Riemann hypothesis is
true for this $L(s, \chi)$ and if $\sigma > \frac{1}{2} + \delta$.   But if $L(\rho, \chi) = 0$
and $\beta > \frac{1}{2}$, then we take $s = \rho$;   the second of our "large"
terms vanishes leaving the three "small" terms to make up a
quantity $e^{-\frac{1}{Y}} \gg 1$.   We obtain our results by estimating how

frequently our "small" terms may be large.

We consider (12.24) with $s = \rho$. We are interested in zeros $\rho = \beta + i\gamma$ with $\beta \geq \sigma$. We first treat the "$\varepsilon(\chi)$" term. As $\varepsilon(\chi) = 1$ only if χ is principal, this term arises only when

$$L(s,\chi) = \zeta(s) \prod_{p|\gamma} (1 - p^{-s}).$$

But if $|M_x(1,\chi_0) \gamma^{1-\rho} \Gamma(1-\rho)| \geq \frac{1}{6}$ then $|\gamma| \ll \log q T$ or $|\gamma| \ll \log QT$, respectively, so this term is $\geq \frac{1}{6}$ for at most $N(\sigma, A_1 \log q T) \ll (\log q T)^2$ or $N(\sigma, A_1 \log QT) \ll (\log QT)^2$ zeros, respectively. In the integral in (12.24) we may restrict the range of integration to the interval $[-Z, Z]$, where $Z = A_2 \log q T$ or $Z = A_2 \log QT$, respectively, with an error of at most $\frac{1}{6}$, provided A_2 is sufficiently large. As for the left hand side of (12.24) we suppose that $Y \geq 6$, so that $e^{-Y} > \frac{5}{6}$. We also restrict the sum to the range $X < n \leq Y^2$, with an error which is $\leq \frac{1}{6}$ provided Y is larger than some absolute constant. Hence from (12.24) we have

$$\left| \sum_{X < n \leq Y^2} a_n \chi(n) n^{-\rho} e^{-\frac{n}{Y}} \right| \geq \frac{1}{6} \tag{12.25}$$

or

$$\left| \int_{-Z}^{Z} L(\tfrac{1}{2} + i\gamma + iu, \chi) M_x(\tfrac{1}{2} + i\gamma + iu, \chi) Y^{\frac{1}{2} - \beta + iu} \Gamma(\tfrac{1}{2} + \rho + iu) \, du \right| \geq \frac{1}{6}, \tag{12.26}$$

or both. Of our zeros ρ with $\beta \geq \sigma$ and $|\gamma| \leq T$ we take a subset $\mathcal{R}$ containing R of them so that if ρ_1 and ρ_2 are zeros of the same L-function then

$$|\gamma_2 - \gamma_1| \geqslant 3Z. \qquad (12.27)$$

Now if χ is a character mod q and $0 < t \leqslant T$ then (see Davenport, §16)

$$N(\tfrac{1}{2}, t+1, \chi) - N(\tfrac{1}{2}, t, \chi) \ll \log q T,$$

so we can choose our R zeros so that

$$\sum_{\chi} N(\sigma, T, \chi) \ll (R+1)(\log q T)^2 \qquad (12.28)$$

or

$$\sum_{q \leqslant Q} \sum_{\chi}^{*} N(\sigma, T, \chi) \ll (R+1)(\log Q T)^2, \qquad (12.29)$$

respectively.   Let $\mathcal{R}_1$ and $\mathcal{R}_2$ be the subsets of $\mathcal{R}$, containing R_1 and R_2 elements, for which (12.25) and (12.26) hold, respectively.   Then $R \leqslant R_1 + R_2$.   We have now reduced the problem to a consideration of situations in which Dirichlet polynomials take on large values at well-spaced points.

We now prove Theorem 12.1.   Our main tools are Theorems 7.6, 8.3, and Corollary 10.4.   If χ is a character mod q, induced by the primitive character χ^*, then $L(s, \chi)$ and $L(s, \chi^*)$ have precisely the same zeros in the half-plane $\sigma > 0$. We consider $L(\sigma, \chi^*)$ instead of $L(s, \chi)$ because in Chapter 10 we gave a mean value theorem not for $\sum_{\chi} |L(s, \chi)|^4$, but only for $\sum_{\chi} |L(s, \chi^*)|^4$.

We now treat $\mathcal{R}_1$, where χ is replaced by χ^* in (12.25).

We divide the sum (12.25) into $< 3 \log Y$ subsums over intervals of the sort $[2^k, 2^{k+1})$. Hence we see that there is a U, $X \leq U \leq Y^2$, such that

$$\left| \sum_{U}^{2U} a_n \chi^*(n) n^{-\rho} e^{-\frac{n}{Y}} \right| \geq (18 \log Y)^{-1} \tag{12.30}$$

for $\gg R_1 (\log Y)^{-1}$ zeros of $\mathcal{R}_1$. Thus from Theorem 7.6 we see that

$$R_1 (\log Y)^{-3} \ll \sum_{X} \sum_{\substack{\rho \in \mathcal{R}_1 \\ L(\rho, \chi) = 0}} \left| \sum_{U}^{2U} a_n \chi^*(n) n^{-\rho} e^{-\frac{n}{Y}} \right|^2$$

$$\ll (\log q T)^4 (q T U^{1-2\sigma} + U^{2-2\sigma}) e^{-\frac{2U}{Y}}$$

$$\ll (\log q T)^4 (q T X^{1-2\sigma} + Y^{2-2\sigma}),$$

so

$$R_1 \ll (q T X^{1-2\sigma} + Y^{2-2\sigma})(\log q T)^7. \tag{12.31}$$

We use this in proving (12.9). To prove (12.10) we apply Theorem 8.3; we take $S(s, \chi)$ to be the sum in (12.30), and we take $\mathcal{L}$ to be the set of those $\rho \in \mathcal{R}_1$ for which (12.30) holds. If

$$X^{2\sigma-1} \geq A_3 (q T)^{\frac{1}{2}} (\log q T)^6 \tag{12.32}$$

where A_3 is sufficiently large, then the term arising from $|\mathcal{L}|$ in (8.8) is less than one half the size of the left hand side of (8.8), and so may be ignored.   From (8.8) we thus have

$$R_1 \ll (\log q T)^5 \, U^{2-2\sigma} e^{-\frac{2U}{Y}} \ll Y^{2-2\sigma} (\log q T)^5. \tag{12.33}$$

We now treat R_2 in two ways;   we assume that (12.26) holds
with χ replaced by χ^*.   For each $\rho \in R_2$ let t_ρ be a real
number satisfying $|\gamma - t_\rho| \le Z$ for which $|L(\tfrac{1}{2}+it_\rho, \chi^*) M_x(\tfrac{1}{2}+it_\rho, \chi^*)|$
is maximum.   As $N(\tfrac{1}{2}, T, \chi^*) \ll T \log q T$ we may assume that
$\beta \ge \sigma \ge \tfrac{1}{2} + (\log q T)^{-1}$.   Hence

$$\int_{-\infty}^{+\infty} |\Gamma(\tfrac{1}{2} - \beta + i u)| \, du \ll \log q T.$$

Thus from (12.26) we have

$$|L(\tfrac{1}{2}+it_\rho, \chi^*) \, M_x(\tfrac{1}{2}+it_\rho, \chi^*)| \gg Y^{\sigma - \frac{1}{2}} (\log q T)^{-1}. \tag{12.34}$$

From (12.27) and the definition of the t_ρ we see that if ρ and
ρ' are zeros of the same L-function then

$$|t_\rho - t_{\rho'}| \ge Z. \tag{12.35}$$

From (12.34) and Hölder's inequality we have

$$R_2 \, Y^{\frac{4}{3}\sigma - \frac{2}{3}} (\log q T)^{-\frac{4}{3}} \ll \sum_{\rho \in R_2} |L(\tfrac{1}{2}+it_\rho, \chi^*) M_x(\tfrac{1}{2}+it_\rho, \chi^*)|^{\frac{4}{3}}$$

$$\le \left(\sum_{\rho \in R_2} |L(\tfrac{1}{2}+it_\rho, \chi^*)|^4 \right)^{\frac{1}{3}} \left(\sum_{\rho \in R_2} |M_x(\tfrac{1}{2}+it_\rho, \chi^*)|^2 \right)^{\frac{2}{3}}.$$

We now apply Corollary 10.4 to the first term, and Theorem 7.6
to the second.   We have

$$R_2 \ll Y^{\frac{2}{3}-\frac{4}{3}\sigma} q T (\log q T)^5,$$

(12.36)

provided $X \leq qT$.

We now give our second treatment of $\mathcal{R}_2$. We follow the above to the inequality (12.35). Let V be a parameter to be chosen later. The $\rho \in \mathcal{R}_2$ for which $|L(\frac{1}{2}+it_\rho, \mathcal{X}^*)| \geq V$ number

$$\ll q T V^{-4} (\log q T)^5,$$

in view of Corollary 10.4. For all other $\rho \in \mathcal{R}_2$ we have, from (12.34),

$$|M_X(\tfrac{1}{2}+it_\rho, \mathcal{X}^*)| \gg Y^{\sigma-\frac{1}{2}} V^{-1} (\log q T)^{-1}.$$

(12.37)

We use Theorem 8.3 to estimate the number of such zeros. We take $M_X(s, \mathcal{X}^*)$ to be our sum in Theorem 8.3, and we take $\mathcal{L}$ to be the set of those $\frac{1}{2}+it_\rho$, $\rho \in \mathcal{R}_2$, for which (12.37) holds. The term arising from $|\mathcal{L}|$ in (8.8) is less than one half the left hand side if

$$Y^{2\sigma-1} \geq A_4 V^2 (qT)^{\frac{1}{2}} (\log q T)^4,$$

(12.38)

where A_4 is some large absolute constant. If (12.38) holds then the $|\mathcal{L}|$ term may be ignored, so from (8.8) we see that the number of $\rho \in \mathcal{R}_2$ for which (12.37) holds is

$$\ll X V^2 Y^{1-2\sigma} (\log q T)^3.$$

Thus if (12.38) holds then

$$R_2 \ll qTV^{-4}(\log_q T)^5 + XY^{1-2\sigma}V^2(\log_q T)^3.$$

(12.39)

We now complete the proof of Theorem 12.1. We take (12.28), (12.31), and (12.36), with $X = qT$ and $Y = (qT)^{\frac{3}{2(2-\sigma)}}$. $Y = (qT)^{\frac{3}{2(2-\sigma)}}$. This gives (12.9). We take (12.28), (12.33), and (12.39) to obtain (12.10). We take X so equality holds in (12.32), we take $V = (qT)^{\frac{3\sigma-2}{4\sigma}}$, and Y so equality holds in (12.38). Then (12.33) and the first term on the right hand side of (12.39) are the same size to within logarithms, and the second term on the right hand side of (12.39) is smaller, provided $\sigma \geqslant \frac{2}{3}$. In fact we could prove (12.10) for $\sigma \geqslant \frac{2}{3}$, but with an inflated power of the logarithm. The bound (12.9) is stronger than (12.10) when $\sigma \geqslant \frac{4}{5}$, so we suppose that $\sigma \geqslant \frac{4}{5}$. An examination of (12.28), (12.33), and (12.39) gives (12.10).

The proof of Theorem 12.2 is the same except that Q^2T replaces qT , (12.29) replaces (12.28), and we use Theorems 7.5, 8.2, and 10.3 instead of Theorems 7.6, 8.3, and Corollary 10.4.

We now prove Theorem 12.3. We may assume that σ is near enough to 1 to ensure that the bound (12.16) is $\leqslant T\log T$, for otherwise the bound is trivial. We follow the preliminaries to the proof of Theorem 12.1 except that we take the contour of integration to the abscissa $\alpha - \sigma$. In treating R_1 we again use the fact that there is a $V,$, $X \leqslant V \leqslant Y^2$, for which

$$\left| \sum_{U}^{2U} a_n \, e^{-\frac{n}{Y}} n^{-\rho} \right| \; \geqslant \; (18 \log Y)^{-1} \tag{12.40}$$

for $\gg R_1 (\log Y)^{-1}$ zeros of $\mathcal{R}_1$. We appeal to Theorem 8.4
with $\theta = \alpha$; the $|\Delta|$ term is negligible if

$$x^{2\sigma - 1 - \alpha} \; \geqslant \; A_1 \, M(\alpha, 8T) \, (\log T)^5,$$

where A_1 is a large absolute constant. In this case

$$R_1 \ll U^{2 - 2\sigma} e^{-\frac{2U}{Y}} (\log T)^6 \ll Y^{2 - 2\sigma} (\log T)^6. \tag{12.41}$$

We now treat $\mathcal{R}_2$. As before, we let t_ρ be a real number
satisfying $|\gamma - t_\rho| \leq Z$ for which

$$|\mathfrak{z}(\alpha + i t_\rho) M_x (\alpha + i t_\rho)|$$

is maximum. We may assume that $\sigma < 1 - (\log T)^{-1}$, so that

$$\int_{-\infty}^{+\infty} |\Gamma(\alpha - \beta + i u)| \, du \ll \log T$$

if $\beta \geq \sigma \geqslant \frac{1 + \alpha}{2}$. Hence from our modified form of (12.26)
we have

$$|\mathfrak{z}(\alpha + i t_\rho) M_x (\tfrac{1}{2} + i t_\rho)| \gg Y^{\sigma - \alpha} (\log T)^{-1}.$$

Now $|t_\rho| \leq 8T$, so

$$|M_x (\tfrac{1}{2} + i t_\rho)| \gg Y^{\sigma - \alpha} (M(\alpha, 8T) \log T)^{-1}.$$

Hence there is a V, $1 \leq V \leq X$, for which

$$\left| \sum_{V}^{2V} \mu(n) \, n^{-\alpha + it_\rho} \right| \gg Y^{\sigma - \alpha} (M(\alpha, 8T))^{-1} (\log T)^{-2}$$

for $\gg R_2 (\log T)^{-1}$ of our zeros. We appeal to Theorem 8.4 with $\theta = \alpha$. If

$$Y^{2\sigma - 2\alpha} > A_2 (M(\alpha, 8T))^3 X^{1-\alpha} (\log T)^4 \tag{12.42}$$

then the $|\Delta|$ term is negligible, and so

$$R_2 \ll X^{2-2\alpha} Y^{2\alpha - 2\sigma} (M(\alpha, 8T))^2 (\log T)^5. \tag{12.43}$$

We take X so equality holds in (12.40) and Y so equality holds in (12.42). Then

$$R_1 \ll \left(A_3 \, M(\alpha, 8T)(\log T)^5 \right)^{\frac{2(1-\sigma)(3\sigma - 1 - 2\alpha)}{(2\sigma - 1 - \alpha)(\sigma - \alpha)}} (\log T)^6,$$

and R_2 is smaller. This, with (12.28), gives (12.16).

To deduce Corollary 12.4 we merely take $\alpha = \frac{1}{2}$, and use the estimate $M(\frac{1}{2}, T) \ll T^{\frac{1}{6}} \log T$ (Titchmarsh's Theorem 5.12). To deduce Corollary 12.5 we take $\alpha = 4\sigma - 3$, and use Richert's bound (11.6).

In order to prove Theorem 12.6 we use Gallagher's Lemma 1.2 to derive a discrete form of Conjecture 9.2.

LEMMA 12.7. _Suppose_ _that_ Conjecture 9.2 _is_ _valid_. _If_

$$S(s) = \sum_{n=N}^{2N} a_n \, n^{-s},$$

if $\mathcal{L}$ is a set of complex numbers $s = \sigma + it$ for which

$$\sigma > \sigma_0 \tag{12.44}$$

and

$$T_0 + \frac{\delta}{2} \le t \le T_0 + T - \frac{\delta}{2} \tag{12.45}$$

for all s in $\mathcal{L}$, and if

$$|t - t'| \ge \delta \tag{12.46}$$

for distinct s and s' in $\mathcal{L}$, then

$$\sum_{s \in \mathcal{L}} |S(s)|^{2\nu} \ll (T + N^\nu)(\delta^{-1} + \log N) \left(\sum_{N^2}^{4N^2} |b_n|^2 n^{-2\sigma_0} \right)^{\frac{\nu}{2}}, \tag{12.47}$$

uniformly for $1 \le \nu \le 2$. Here $b_n = \sum_{d|n} |a_d a_{\frac{n}{d}}|$.

To prove this we take $f(t) = S(\sigma_0 + it)^{2\nu}$ in Lemma 1.2. Then (1.4) gives

$$\sum_{s \in \mathcal{L}} |S(\sigma_0 + it)|^{2\nu} \le \delta^{-1} \int_{T_0}^{T_0+T} |S(\sigma_0 + iu)|^{2\nu} du + 2 \int_{T_0}^{T_0+T} |S(\sigma_0 + iu) S'(\sigma_0 + iu)| du.$$

By Hölder's inequality the second integral is

$$\le \left(\int_{T_0}^{T_0+T} |S(\sigma_0 + iu)|^{2\nu} du \right)^{\frac{2\nu-1}{2\nu}} \left(\int_{T_0}^{T_0+T} |S'(\sigma_0 + iu)|^{2\nu} du \right)^{\frac{1}{2\nu}}.$$

Hence if (9.9) holds then

$$\sum_{s \in \mathcal{A}} |S(\sigma_0 + it)|^{2\nu} \ll (T + N^\nu)(\delta^{-1} + \log N)\left(\sum_{N^2}^{4N^2} |b_n|^2 \, n^{-2\sigma_0}\right)^{\frac{\nu}{2}}. \qquad (12.48)$$

We now deduce (12.46) from this in the same way we deduced Theorem 7.5 from Theorem 7.3. We write

$$S(s, v) = \sum_{N \leq n \leq v} a_n n^{-s},$$

so that

$$S(s) = S(\sigma_0 + it)(2N)^{\sigma_0 - \sigma} + (\sigma - \sigma_0)\int_N^{2N} S(\sigma_0 + it, v) v^{-\sigma + \sigma_0 - 1} dv$$

$$\ll |S(\sigma_0 + it)| + N^{-1}\int_N^{2N} |S(\sigma_0 + it, v)| dv.$$

Hence

$$|S(s)|^{2\nu} \ll |S(\sigma_0 + it)|^{2\nu} + N^{-1}\int_N^{2N} |S(\sigma_0 + it)|^{2\nu} dv.$$

This, with (12.47), gives (12.46).

We now prove Theorem 12.6. We take $X = T^{\frac{1}{2}}$, $Y = T$, and follow the preliminary portion of the proof of Theorem 12.1. We first treat $\mathcal{R}_1$. Again there is a U, $T^{\frac{1}{2}} \leq U \leq T^2$, for which

$$\left|\sum_{n=U}^{2U} a_n n^{-s} e^{-\frac{n}{T}}\right| \geq (18 \log T)^{-1}$$

for $\gg R_1 (\log T)^{-1}$ of our zeros in R_1. If $T^{\frac{1}{2}} \le U \le T$ then we define ν by the relation $U^{\nu} = T$. Then by Lemma 12.7 we have

$$R_1 \ll T(\log T)^{2+2\nu} \left(U^{2-4\sigma}(\log U)^8 \right)^{\frac{\nu}{2}}$$

$$\ll T^{2-2\sigma}(\log T)^{14}.$$

On the other hand if $T < U \le T^2$ then we apply Theorem 7.5. We have

$$R_1 \ll U^{2-2\sigma}(\log T)^8 e^{-\frac{2\nu}{T}}$$

$$\ll T^{2-2\sigma}(\log T)^8.$$

Thus in either case we have

$$R_1 \ll T^{2-2\sigma}(\log T)^{14}. \tag{12.49}$$

We treat R_2 as before so that we have well-spaced t_ρ for which $|\Im(\frac{1}{2}+it_\rho) M_x(\frac{1}{2}+it_\rho)| \gg T^{\sigma-\frac{1}{2}}(\log T)^{-1}$.

Hence

$$R_2 T^{2\sigma-1}(\log T)^{-2} \ll \sum_{\rho \in R_2} |\Im(\frac{1}{2}+it_\rho) M_x(\frac{1}{2}+it_\rho)|^2.$$

By Cauchy's inequality this is

$$\le \left(\sum_{\rho \in R_2} |\Im(\frac{1}{2}+it_\rho)|^4 \right)^{\frac{1}{2}} \left(\sum_{\rho \in R_2} |M_x(\frac{1}{2}+it_\rho)|^4 \right)^{\frac{1}{2}}.$$

From Theorems 10.3 and 7.3 we see that this is

$$\ll \left(T(\log T)^5\right)^{\frac{1}{2}} \left(T(\log T)^5\right)^{\frac{1}{2}}.$$

Hence

$$R_2 \ll T^{2-2\sigma}(\log T)^7.$$

This, with (12.49) and (12.28), gives (12.22) as required.

Least character non-residues and arg L($\frac{1}{2}$ + it, χ)

Many results are known which connect bounds for
a) the size of $L(1, \chi)$; b) the size of the character
sums $\sum_{n \leq x} \chi(n)$; c) the size of the least n for which
$\chi(n) \neq 1$, $\chi(n) \neq 0$; d) the size of $L(s, \chi)$ in the
critical strip; e) the width of the zero-free region for
$L(s, \chi)$; and f) the size of $N(\sigma, T, \chi)$. In this
chapter we discuss those hypotheses which allow one to say
something concerning the size of the least n for which
$\chi(n) \neq 1$, $\chi(n) \neq 0$. We denote this least n by n_χ .

If χ is a character mod p and

$$\left| \sum_{n \leq N} \chi(n) \right| < N$$

for some $N < p$ then it is obvious that $n_\chi \leq N$. Moreover a
simple argument of I.M. Vinogradov shows that if χ is a
character mod p and if

$$\left| \sum_{n \leq N} \chi(n) \right| < \delta N$$

then $n_\chi < N^{e^{-\frac{1}{2}} + \varepsilon}$, where $\varepsilon = \varepsilon(\delta, N)$. Hence from the Pólya -
Vinogradov inequality [151] , [221] (see also [105] , [186]),
we have

$$n_\chi < p^{\frac{1}{2\sqrt{e}} + \varepsilon} ,$$

and from Burgess's [24] , [25] , [27] bounds we have

$$n_\chi < p^{\frac{1}{4\sqrt{e}} + \varepsilon}.$$

The generalized Lindelöf hypothesis states that

$$L(\tfrac{1}{2} + it, \chi) \ll (q(|t|+2))^\varepsilon. \tag{13.1}$$

from this it would follow that

$$\sum_{n \leq N} \chi(n) \ll N^{\frac{1}{2}} q^\varepsilon, \tag{13.2}$$

and hence, without need of Vinogradov's argument, we would have

$$n_\chi \ll q^\varepsilon. \tag{13.3}$$

In the other direction, from (13.2) one has $L(\tfrac{1}{2}, \chi) \ll q^\varepsilon$, and

$$L(1, \chi) = o(\log q). \tag{13.4}$$

Unconditionally we have [26] , [28] only $L(\tfrac{1}{2}, \chi) < q^c$, $c > 0$, and $L(1, \chi) \ll \log q$.

The hypothetical bounds (13.1), (13.2), (13.3), (13.4) can be deduced from hypothetical statements concerning the distribution of zeros of $L(s, \chi)$. For example, (13.1) is equivalent (see Titchmarsh, Theorem 13.5) to the assertion that

$$N(\sigma, T+1, \chi) - N(\sigma, T, \chi) = o_\sigma(\log qT)$$

for $T \geq 1$ and $\sigma > \tfrac{1}{2}$. Linnik [129] proved that (13.3) holds if

$$N(1-\delta, t+\delta, \chi) - N(1-\delta, t, \chi) = o(\delta \log q) \qquad (13.5)$$

when $0 \le t \le 1$ and $0 \le \delta \le \Delta$ and Δ is some absolute constant,
One consequence of Linnik's hypothesis (13.5) is that if
$L(\rho, \chi) = 0$ then

$$|1 - \rho| \log q \to \infty. \qquad (13.6)$$

On the other hand, the best known zero-free region when $0 \le |t| \le q$
is

$$\sigma \le 1 - c(\log q)^{-1}.$$

Rodosskiĭ [173] has shown that the consequence (13.6) of Linnik's
hypothesis is still enough to imply (13.3). He showed that if
$L(s, \chi) \ne 0$ in the rectangles

$$1 - \frac{\psi}{\log q} \le \sigma \le 1, \quad 0 < |t| \le \min(1, e^{\psi}(\log q)^{-1}),$$

where $e \le \psi \le \frac{1}{2} \log q$, then

$$n_\chi < q^{A \psi^{-1} \log \psi},$$

where A is a positive absolute constant. If ψ can be taken
large for $q > q_0(\psi)$ then (13.3) follows. In view of our
results of Chapter 11 we can say that (13.6) and hence (13.3)
would follow if we could establish the following statement:

Let α, β be fixed positive numbers, $1 \ge \alpha > \beta > 0$; for
any $\varepsilon > 0$ there exist δ and q_0 such that

$$N(1-\delta h, t+h, \chi) - N(1-\delta h, t, \chi) < \varepsilon h \log q, \tag{13.7}$$

whenever $0 < t \leq 1$, $q > q_0$, and $(\log q)^{-\alpha} \leq h \leq (\log q)^{-\beta}$.

The above is weaker than the hypothesis (13.5). It is known (see Prachar, Chapter 10, Lemma 2.1) that

$$N(1-\delta, t+\delta, \chi) - N(1-\delta, t, \chi) \ll \delta \log q$$

for $0 \leq \delta \leq 1$, $0 \leq t \leq 1$. We see that (13.5) hypothesizes a sharper bound for a narrower rectangle.

We note that n_χ must be a prime, for if $n_\chi = ab$ then one of $\chi(a)$ and $\chi(b)$ is different from 1. This suggests that we might ask for bounds for the least prime p for which $\chi(p) = 1$. If the order of χ is bounded then this problem is similar to the problem of bounding n_χ. One difference, however, is that if χ is a real character then the possibility of the existence of a real zero near to 1 must be excluded; for n_χ such a zero always has a beneficial effect.

The Riemann hypothesis for $L(s, \chi)$ implies all the other hypotheses that we have discussed above. In particular, Ankeny [1] proved a special case of

THEOREM 13.1 (Ankeny). <u>If</u> χ <u>is a character</u> mod q <u>and</u> $L(s, \chi)$ <u>has no zero with real part greater than</u> $\frac{1}{2}$, <u>then</u> <u>there is an integer</u> n <u>with</u> $\chi(n) \neq 1$, $\chi(n) \neq 0$, <u>for which</u>

$$1 < n \ll (\log q)^2. \tag{13.8}$$

Ankeny's analysis is elaborate, and uses (among other

things) a theorem of Selberg [190] which asserts that if the Riemann hypothesis is true for $L(s, \chi)$ then

$$S(t, \chi) \ll \frac{\log q \tau}{\log \log q \tau},$$ (13.9)

where $\tau = |t| + 2$, and $S(t, \chi) = \frac{1}{\pi} \arg L(\frac{1}{2} + it, \chi)$ is defined by continuous variation from $+\infty + it$. In addition to giving a simple proof of Ankeny's theorem, we give a simple proof of the following related result:

THEOREM 13.2. Let χ be a character mod q, and let d be the order of χ. Suppose that the Riemann hypothesis is true for the L-functions $L(s, \chi^k)$, $1 \leq k \leq d-1$. Then for any d-th root of unity η, there is a prime number p with $\chi(p) = \eta$ and

$$p \ll d (\log q)^2.$$ (13.10)

By elaborating on Ankeny's argument we also derive

THEOREM 13.3. If χ is a primitive character mod q and $L(s, \chi)$ has no zeros with real part greater than $\frac{1}{2}$, then

$$\sum_{n=1}^{\infty} \Lambda(n) \chi(n) \left(e^{-\frac{n}{2x}} - e^{-\frac{n}{x}} \right) \ll x^{\frac{1}{2}} M(\chi) \log x + x^{-1} \log q$$ (13.11)

for $x \geq 1$, where

$$M(\chi) = \int_{-\infty}^{+\infty} |S(u, \chi)| e^{-|u|} du$$ (13.12)

and $S(t, \chi)$ is as defined above.

From this we have immediately

COROLLARY 13.4.   Under the above hypotheses

$$n_\chi \ll (M(x) \log\log q)^2 + (\log q)^{\frac{1}{2}}.$$ (13.13)

In the opposite direction Fridlender [66] and Salié [181] independently showed that there are primitive characters χ for which

$$n_\chi = \Omega(\log q).$$ (13.14)

From this and (13.13) we see that if the generalized Riemann hypothesis is valid then

$$S(t,\chi) = \Omega((\log q)^{\frac{1}{2}}(\log\log q)^{-1})$$

with $|t| \leq \log\log q$.   This is essentially a q-analogue of a result of Bohr and Landau [12] (see also [189]).   If we assume the generalized Riemann hypothesis then we can improve on (13.14); we have

THEOREM 13.5.   If the Riemann hypothesis is true for all L-functions of real characters χ , then for real primitive characters we have

$$n_\chi = \Omega((\log q)(\log\log q)).$$ (13.15)

This, with Corollary 13.4 gives

COROLLARY 13.6.   If the Riemann hypothesis is true for all L-functions of a real character then there are pairs of real primitive characters χ and real numbers t with

$$| S(t, \chi)| \gg (\log q)^{\frac{1}{2}} (\log\log q)^{-\frac{1}{2}}, \qquad (13.16)$$

where q <u>is arbitrarily large and</u> $|t| \leq \log\log q$

These results have their t - analogues; for the zeta function we prove

THEOREM 13.17. <u>If the Riemann hypothesis is valid then</u>

$$S(t) = \Omega \left((\log t)^{\frac{1}{2}} (\log\log t)^{-\frac{1}{2}} \right). \qquad (13.17)$$

Previously Bohr and Landau [12] proved that

$$S(t) = \Omega_{\pm} \left((\log t)^{\frac{1}{2} - \epsilon} \right);$$

Selberg [189] observed that their method gives

$$S(t) = \Omega_{\pm} \left((\log t)^{\frac{1}{2}} (\log\log t)^{-1} \right).$$

By modifying our approach we could replace Ω in (13.17) by $\Omega_{\pm}$.
We could also show that

$$S(t) - S(t + (\log\log t)^{-1}) = \Omega_{\pm} \left((\log t)^{\frac{1}{2}} (\log\log t)^{-\frac{1}{2}} \right). \qquad (13.18)$$

We now prove Theorem 13.1 and 13.2; for the moment we assume that χ is primitive. We have the easily proved formula

$$\sum_{n \leq N} (1 - \tfrac{n}{N}) \Lambda(n) \chi(n) = - \sum_{\rho} \frac{N^{\rho}}{\rho(\rho+1)} + O(\log q N).$$

If we restrict the sum on the left to primes then we alter the quantity by an amount that is $\ll N^{\frac{1}{2}}$. If we drop the condition that χ is primitive then we alter the left hand side by an amount which is $\ll \log q$. Hence for any non-principal character we have

$$\sum_{p \leq N} \left(1 - \frac{p}{N}\right)(\log p)\, \chi(p) = -\sum_{\rho} \frac{N^{\rho}}{\rho(\rho+1)} + O(N^{\frac{1}{2}}) + O(\log q).$$

But $N(0, t+2, \chi) - N(0, t, \chi) \ll \log q t$ for $t \geq 2$, so the sum over zeros is absolutely convergent, to a value $\ll N^{\frac{1}{2}} \log q$ if $\beta \leq \frac{1}{2}$ for all zeros ρ. Hence

$$\sum_{p \leq N} \left(1 - \frac{p}{N}\right)(\log p)\, \chi(p) \ll N^{\frac{1}{2}} \log q. \tag{13.19}$$

If $\chi(p) = 1$ or 0 for all $p \leq N$ then by the prime number theorem the left hand side is

$$= \frac{1}{2} N + o(N) + O(\log q),$$

from which we have $N \ll N^{\frac{1}{2}} \log q$, and so Theorem 13.1 follows.

To prove Theorem 13.2 we note that

$$d^{-1} \sum_{k=1}^{d} \eta^{-k} \chi^{k}(n) = \begin{cases} 1 & \text{if } \chi(n) = \eta, \\ 0 & \text{otherwise.} \end{cases}$$

We take (13.19) for the characters χ^{k}, $1 \leq k \leq d-1$, and use the prime number theorem for $\chi^{d} = \chi_{o}$. We obtain

$$\sum_{\substack{p \leq N \\ \chi(p) = \eta}} \left(1 - \frac{p}{N}\right) \log p = \frac{N}{2d} + o\left(\frac{N}{d}\right) + O(N^{\frac{1}{2}} \log q).$$

If the left hand side is zero then $\frac{N}{d} \ll N^{\frac{1}{2}} \log q$, and the result follows.

Ankeny [1] proved that if the Riemann hypothesis is true for $L(s, \chi)$ then

$$\sum_{\rho} x^{\rho} \Gamma(\rho) \ll x^{\frac{1}{2}} (\log x)(\log q)(\log\log q)^{-1} + x^{\frac{1}{2}} (\log q)(\log x)^{-1},$$

where ρ runs over all zeros of $L(s, \chi)$ on $\sigma = \frac{1}{2}$. By refining his analysis we could show that under the same hypotheses

$$\sum_{\rho} x^{\rho} \Gamma(\rho) \ll x^{\frac{1}{2}} (\log x) M(\chi) + \log q x.$$

For use in the proofs of Theorems 13.3 and 13.7 we establish

LEMMA 13.8. Let χ be a primitive character mod γ , and suppose that $L(s, \chi)$ has no zeros with real part greater than $\frac{1}{2}$. Then

$$\sum_{\rho} \left((2x)^{\rho - it} - x^{\rho - it} \right) \Gamma(\rho - it) \ll x^{\frac{1}{2}} M(t, \chi) \log x$$
$$+ x^{-1} \log q \qquad (13.20)$$

for $x > 1$, where

$$M(t, \chi) = \int_{-\infty}^{+\infty} |S(u+t, \chi)| e^{-|u|} du. \qquad (13.21)$$

To prove this lemma we let

$$N(u, \chi) = \frac{u}{2\pi} \log \frac{q}{2\pi} + \frac{1}{2\pi} \arg \Gamma(\tfrac{1}{4}+iu) - \frac{\chi(-1)}{8} + S(u,\chi) - S(0,\chi),$$

so that $N(u,\chi)$ counts zeros of $L(s,\chi)$. Hence the left hand side of (13.20) is

$$= \int_{-\infty}^{+\infty} ((2x)^{\frac{1}{2}+iu} - x^{\frac{1}{2}+iu}) \Gamma(\tfrac{1}{2}+iu-it) \, d\,N(u,\chi).$$

We write this as $\int_{1} + \int_{2}$, where

$$\int_{1} = \int_{-\infty}^{+\infty} ((2x)^{\frac{1}{2}+iu} - x^{\frac{1}{2}+iu}) \Gamma(\tfrac{1}{2}+iu-it) \left(\frac{1}{2\pi} \log \frac{q}{2\pi} + \frac{1}{4\pi} \frac{\Gamma'}{\Gamma}(\tfrac{1}{4}-iu) + \frac{1}{4\pi} \frac{\Gamma'}{\Gamma}(\tfrac{1}{4}+iu) \right) du,$$

and

$$\int_{2} = \int_{-\infty}^{+\infty} ((2x)^{\frac{1}{2}+iu} - x^{\frac{1}{2}+iu}) \Gamma(\tfrac{1}{2}+iu-it) \, d\,S(u,\chi).$$

Now

$$\int_{1} = \frac{1}{2\pi i} \int_{\frac{1}{2}-i\infty}^{\frac{1}{2}+i\infty} ((2x)^{w-it} - x^{w-it}) \Gamma(w-it) \left(\log \frac{q}{2\pi} + \tfrac{1}{2}\Gamma(w) + \tfrac{1}{2}\Gamma(1-w) \right) dw.$$

We take the $\log \frac{q}{2\pi}$ term to the abscissa $\mathrm{Re}\, w = -\frac{3}{2}$; we pass a pole at $w = -1+it$ with residue $\ll x^{-1} \log 2q$, and the new integral is $\ll x^{-\frac{3}{2}} \log 2q$. The $\Gamma(w)$ and $\Gamma(1-w)$ terms we take to the abscissa $\mathrm{Re}\, w = -\frac{1}{2}$. The function $\Gamma(w)$ has a pole at $w = 0$, which gives a residue $\ll e^{-|t|} \ll 1$; the new integral is $\ll x^{-\frac{1}{2}}$.

On the other hand

$$\int_{2} = \int_{-\infty}^{+\infty} S(u,x) \frac{d}{du}\left(\left((2x)^{\frac{1}{2}+iu-it} - x^{\frac{1}{2}+iu-it}\right)\Gamma(\tfrac{1}{2}+iu-it)\right) du$$

$$\ll \int_{-\infty}^{+\infty} |S(u,x)| \, x^{\frac{1}{2}}(\log x)\, e^{-|u-t|} du.$$

Hence the left hand side of (13.20) is

$$\ll x^{\frac{1}{2}}(\log x) M(t,x) + x^{-1}\log 2q + 1.$$

Now $M(t,x) \gg 1$ in any case, so the $x^{-1}\log 2q + 1$ can be discarded. Hence we have (13.20).

We now prove Theorem 13.3. If χ is a primitive character modulo $q > 1$ then

$$\sum_{n=1}^{\infty} \Lambda(n)\, \chi(n)\, n^{-it}\left(e^{-\frac{n}{2x}} - e^{-\frac{n}{x}}\right) = \frac{1}{2\pi i}\int_{2-i\infty}^{2+i\infty} -\frac{L'}{L}(w+it,\chi)\, \Gamma(w)\left((2x)^{w} - x^{w}\right) dw$$

$$= - \sum_{\rho} \left((2x)^{\rho-it} - x^{\rho-it}\right)\Gamma(\rho-it)$$

$$+ O(1). \tag{13.22}$$

From Lemma 13.8 with $t = 0$ we have

$$\sum_{n=1}^{\infty} \Lambda(n)\chi(n)\left(e^{-\frac{n}{2x}} - e^{-\frac{n}{x}}\right) \ll x^{\frac{1}{2}}(\log x) M(x) + x^{-1}\log q,$$

which is (13.11).

We now prove Theorem 13.5. Let P_y be the set of primes p for which $\left(\frac{p_1}{p}\right) = 1$ for all primes $p_1 \leq y$. Then

$$\sum_{\substack{x < p \leq 2x \\ p \in P_y}} \log p = 2^{-\pi(y)} \sum_{x < p \leq 2x} (\log p) \prod_{p_1 \leq y} \left(1 + \left(\frac{p_1}{p}\right)\right)$$

$$= 2^{-\pi(y)} \sum_m \psi(2x, \chi_m) - \psi(x, \chi_m)$$

for $x > y$, where χ_m is the character determined by quadratic reciprocity so that

$$\chi_m(p) = \prod_{p_1 | m} \left(\frac{p_1}{p}\right).$$

Now the generalized Riemann hypothesis implies that $\psi(x, \chi) \ll x^{\frac{1}{2}} (\log qx)^2$, so

$$\sum_{\substack{x < p \leq 2x \\ p \in P_y}} \log p = 2^{-\pi(y)} x + O\left(x^{\frac{1}{2}} y^2 + x^{\frac{1}{2}} (\log x)^2\right).$$

Hence if $y = c (\log x)(\log\log x)$ where c is a small positive constant, then the right hand side above is positive as x tends to infinity. Hence for $x > x_0$ there is a $p \in P_y$ with y as above. This proves the result.

We now prove Theorem 13.7. For the zeta function we have (13.22), except that there is an additional term $x^{1-it} \Gamma(1-it)$ on the right, due to the pole of $\zeta(s)$ at $s = 1$. From Lemma 13.8 with $q = 1$ we have

$$\sum_{n=1}^{\infty} \Lambda(n)\, n^{-it}\left(e^{-\frac{n}{2x}} - e^{-\frac{n}{x}}\right) \ \ll\ x^{\frac{1}{2}}(\log x)\int_{-\infty}^{+\infty} |S(u+t)|\, e^{-|u|}\, du$$

$$+\, x\, e^{-|t|}$$

By Dirichlet's theorem we see that for any large T there is a t for which $T^{\frac{1}{2}} \leqslant t \leqslant T$ and

$$|n^{-it} - 1| \leqslant \tfrac{1}{2}$$

for all $n \leqslant N = c(\log T)(\log\log T)$ for which $\Lambda(n) \neq 0$, provided c is a sufficiently small positive constant. We take $x = \frac{N}{16}$. Then

$$\left| \sum_{n=1}^{\infty} \Lambda(n)\, n^{-it}\left(e^{-\frac{n}{2x}} - e^{-\frac{n}{x}}\right) \right| \ \gg\ x,$$

so

$$\int_{-\infty}^{+\infty} |S(u+t)|\, e^{-|u|}\, du \ \gg\ (\log t)^{\frac{1}{2}}(\log\log t)^{-\frac{1}{2}},$$

which gives (13.17).

The prime number theorems of Hoheisel and Selberg

In 1930 Hoheisel $[88]$ proved that there is a prime in an interval of the form

$$[x, x + x^{\theta+\epsilon}] \tag{14.1}$$

where $\theta = 1 - (3300)^{-1}$ and $x > x_o(\epsilon)$. Heilbronn $[84]$ then improved some estimates to show that one can take $\theta = 1 - (250)^{-1}$ in (14.1). Both Hoheisel and Heilbronn remarked that if the zero-free region for the zeta function could be widened, then one could take $\theta = \frac{3}{4}$ in (14.1). Čudakov $[41]$ later succeeded in doing this. Ingham $[97]$ and Fogels $[63]$, $[64]$ then described ways in which good values of θ might be obtained. In particular, Ingham showed that if

$$N(\sigma, T) \ll T^{\alpha(1-\sigma)} (\log T)^A \tag{14.2}$$

uniformly for $\frac{1}{2} \leq \sigma \leq 1$, then one may take $\theta = 1 - \alpha^{-1}$ in (14.1). He also proved (12.3), so that one may take

$$\theta = (1+4c)(2+4c)^{-1}$$

if $\zeta(\frac{1}{2} + it) \ll t^c \log t$ for $t > 2$. From $c = \frac{1}{6}$ Ingham obtained $\theta = \frac{5}{8}$, and a slightly better result followed from Walfisz's $[225]$ estimate for c . Later Titchmarsh $[204]$ gave $c = \frac{19}{116}$, from which one obtains $\theta = \frac{48}{77} = \frac{5}{8} - \frac{1}{616}$. Then Min $[139]$ gave $c = \frac{15}{92} + \epsilon$, so that $\theta = \frac{38}{61} = \frac{5}{8} - \frac{1}{488}$. Finally Haneke $[79]$ gave $c = \frac{6}{37}$, from which one has $\theta = \frac{61}{98} = \frac{5}{8} - \frac{1}{392}$.

From Theorem 12.1 with $q = 1$ or Theorem 12.2 with $Q = 1$ we see that we have (14.2) with $\alpha = \frac{6}{5}$ and $A = 12$. Hence without using a bound for $|\zeta(\frac{1}{2} + it)|$ we have

THEOREM 14.1. <u>For any</u> $\varepsilon > 0$ <u>and any</u> $x > x_0(\varepsilon)$ <u>there is a prime in the interval</u>

$$[x, \ x + x^{\frac{2}{5} + \varepsilon}]. \tag{14.3}$$

Similar results may be derived for primes in arithmetic progressions (see Jutila [101]). From the density hypothesis (12.17) we would have $\theta = \frac{1}{2}$ in (14.1), while Cramér [37] showed that from the Riemann hypothesis it follows that there is a prime in the interval $[x, x + C x^{\frac{1}{2}} \log x]$, where C is a large absolute constant. On the other hand Pilz conjectured that one may take $\theta = 0$ in (14.1). Later Cramér [40] conjectured that

$$\varlimsup_{n} \ \frac{p_{n+1} - p_n}{(\log p_n)^2} = 1. \tag{14.4}$$

Shanks [195] has further strengthened this statement, and has produced numerical evidence in support of these conjectures.

Cramér [38], [39] investigated the frequency with which large gaps occur between prime numbers. Selberg [188] sharpened these results. He also proved that (essentially) if (14.2) holds then (14.1) holds for almost all x , provided that $\theta > 1 - 2 \alpha^{-1}$. Hence from (12.3) we see that (14.1) holds for almost all x with

$$\theta = 2c(1+2c)^{-1}.$$

From Theorem 12.1 and (14.2) we have

THEOREM 14.2   <u>Suppose that</u> ε <u>is positive.   There is a</u>
<u>prime in the interval</u>

$$\left[n, n + n^{\frac{1}{3} + \varepsilon} \right] \tag{14.5}$$

<u>for</u> $n \leq N$, <u>with the exception of</u> $o_\varepsilon(N)$ <u>values of</u> n .

Selberg [188] has shown that if the Riemann hypothesis
is true then for almost all x the interval $[x, x + f(x)(\log x)^2]$
contains a prime, where $f(x)$ is any function which tends to
infinity with x .

CHAPTER 15

The Bombieri - Vinogradov theorem
and other applications of the large sieve

A detailed review of applications of the large sieve is
found in the survey article of Barban [7], so we restrict
our attention to more recent work, some of which we have already
mentioned in previous chapters.   Burgess and Elliott [30]
have used the large sieve to give a bound for the average of
the least primitive root.   Joshi [100] improved the earlier
results of Bateman, Chowla, and Erdős [8] concerning the
size of $L(1,\chi)$.   Elliott [55] has shown that the Turán -
Kubilius inequality may be derived from the large sieve.   By
far the most important application has been that of Bombieri
[15] in proving the Bombieri - Vinogradov theorem.
Vinogradov [220] did not make use of the large sieve in
deriving his (slightly weaker) result, but Bombieri's approach
depends on the zero-density estimate (12.12), which in turn
was derived from the large sieve.   Let

$$E(x,q) = \max_{\substack{a \\ (a,q)=1}} \left| \psi(x;q,a) - \frac{x}{\phi(q)} \right|, \qquad (15.1)$$

and let

$$E^*(x,q) = \max_{y \le x} E(y,q). \qquad (15.2)$$

In this notation Bombieri's result is that for any fixed
$A > 0$

$$\sum_{q \leq Q} E^{*}(x,q) \ll x (\log x)^{-A} \qquad (15.3)$$

provided that $Q \leq x^{\frac{1}{2}}(\log x)^{-B}$, where $B = B \cdot 34 + 23$. In his book Davenport gives $B = 4A + 40$. Gallagher [69] has given a simple proof, which gives $B = 16A + 103$. We follow Davenport's redaction, but use Theorem 12.2 in place of (12.12), to prove

THEOREM 15.1. <u>Let</u> $E(x,q)$ <u>and</u> $E^{*}(x,q)$ <u>be defined by</u> (15.1) <u>and</u> (15.2). <u>Then for any</u> $C > 0$

$$\sum_{q \leq Q} E^{*}(x,q) \ll x^{\frac{1}{2}} Q (\log x)^{13} \qquad (15.4)$$

<u>for</u> $x^{\frac{1}{2}}(\log x)^{-C} \leq Q \leq x^{\frac{1}{2}}$, <u>provided</u> $x > x_{0}(C)$.

This corresponds to $B = A + 13$ in (15.3). The main advantage of the Bombieri - Vinogradov theorem is that in some situations it can be used in place of the generalized Riemann hypothesis (GRH). This has been particularly true of problems which were solved assuming the GRH, and then solved unconditionally by Linnik's dispersion method. For example, Titchmarsh [200] proved that the GRH implies that

$$\sum_{\ell < p \leq x} d(p - \ell) \sim x \left(\prod_{p | \ell} (1 - \frac{1}{p}) \right) \left(\prod_{r} (1 + \frac{1}{p(p-1)}) \right)$$

as x tends to infinity, for fixed ℓ. Iseki [99] gave a similar treatment of the conjugate problem, and Hooley [89] did the same for the more difficult problem of Hardy and Littlewood concerning the representation of an integer as a sum of a prime and two squares. These problems were first solved by Linnik (see his book) and others using the dispersion method, but now one can return to the earlier (and simpler) treatments, using the Bombieri - Vinogradov theorem (see Rodriguez [174], and Elliott and Halberstam [57]). Wilson [227] has extended the Bombieri - Vinogradov theorem to algebraic number fields, and Turán has suggested that in this way one might give an unconditional treatment of the work of Hooley [90] on Artin's conjecture.

The Bombieri - Vinogradov theorem has also been used by Bombieri and Davenport [19] , who obtained a new upper bound for

$$E = \lim \frac{p_{n+1} - p_n}{\log p_n}.$$

Bombieri has remarked that their paper could be simplified by further appeals to (15.3), and Huxley [92] made substantial simplifications in the course of deriving further results. Halberstam, Jurkat, and Richert [76] have used the Bombieri - Vinogradov theorem to improve certain results obtained by Selberg's method. In particular they proved that any large positive even integer $2n$ may be written $2n = p + P_3$, where P_k denotes a number which has at most k prime factors. Davenport and Halberstam [48] used the large sieve to strengthen a result of Barban [5] , [6] ; they proved that

$$\sum_{q \leq Q} \sum_{a=1}^{q} {}^{*} (\psi(x; q, a) - \frac{x}{\phi(q)})^2 \ll Q x (\log x)^5, \tag{15.5}$$

provided $x(\log x)^{-A} \leq Q \leq x$. This is weaker than (15.4) in that it bounds a mean square instead of a maximum; on the other hand it is stronger than (15.4) in that it gives a good bound for larger values of Q. We devote Chapter 17 to a discussion of (15.5) and the literature concerning it.

The following is an alternate form of the Bombieri - Vinogradov theorem; this form is very useful in certain applications.

THEOREM 15.2. Let $E(x,q)$ and $E^*(x,q)$ be defined by (15.1) and (15.2). Then for any $C > 0$

$$\sum_{q \leq Q} q\left(E^*(x,q)\right)^2 \ll x^{\frac{3}{2}} Q (\log x)^{14} \tag{15.6}$$

for $x^{\frac{1}{2}}(\log x)^{-C} \leq Q \leq x^{\frac{1}{2}}$, provided $x > x_0(C)$.

The GRH gives

$$E^*(x,q) \ll x^{\frac{1}{2}}(\log qx)^2, \tag{15.7}$$

and we see that (15.4) and (15.6) both follow from this. On the other hand the trivial estimate for $E^*(x,q)$ is

$$E^*(x,q) \ll \left(1 + \frac{x}{q}\right)\log x, \tag{15.8}$$

so (15.4), (15.5), and (15.6) are only a power of a logarithm smaller than trivial. We may expect that

$$E^*(x,q) \ll \left(\frac{x}{q}\right)^{\frac{1}{2}+\varepsilon}\log x \tag{15.9}$$

for $x > q$. This observation has led Halberstam to conjecture that

$$\sum_{q \le Q} E^*(x,q) \ll x (\log x)^{-A} \qquad (15.10)$$

even when $Q = x^{1-\varepsilon}$. Elliott [54] has given a measure-theoretic argument which indicates that is is probable that (15.10) holds for $Q = x (\log x)^{-B}$, $B = B(A)$. Halberstam has informed me that from (15.10) with $Q = x^{\frac{4}{7}}$ it follows that every large even integer $2n$ may be written $2n = p + P_2$. Moreover, Porter (unpublished) has found that one can give an asymptotic formula for

$$\sum_{p \le x} d_k(p-1)$$

if (15.10) holds with $Q = x^{1-\frac{1}{k}}$. We note that (15.5) indicates that (15.9) is true on average, which gives us further hope that (15.10) may hold with $Q = x^{1-\varepsilon}$.

While it seems that the Bombieri - Vinogradov theorem is far from best possible, it is deduced from the following character sum estimate, which is almost certainly essentially best possible.

THEOREM 15.3. For $x^{\frac{1}{2}}(\log x)^{-C} \le Q \le x^{\frac{1}{2}}$ we have

$$\sum_{q \le Q} \frac{1}{\phi(q)} \sum_{\chi \ne \chi_p} \max_{y \le x} |\psi(y,\chi)| \ll x^{\frac{1}{2}} Q (\log x)^{13}. \qquad (15.11)$$

We now show that Theorem 15.1 follows from the above.

As

$$\Psi(y; q, a) = \frac{1}{\phi(q)} \sum_{\chi} \overline{\chi}(a) \Psi(y, \chi), \qquad (15.12)$$

we see that

$$\left| \Psi(y; q, a) - \frac{y}{\phi(q)} \right| \leq \frac{|\Psi(y, \chi_0) - y|}{\phi(q)} + \frac{1}{\phi(q)} \sum_{\chi \neq \chi_0} |\Psi(y, \chi)|. \qquad (15.13)$$

By the prime number theorem the first term is
$$\ll \phi(q)^{-1} y (\log y)^{-c-1} + \phi(q)^{-1} (\log qy)^2 \qquad , \text{ so}$$

$$E^*(x, q) \ll \phi(q)^{-1} x (\log x)^{-c-1} + \phi(q)^{-1} \sum_{\chi \neq \chi_0} \max_{y \leq x} |\Psi(y, \chi)|.$$

Now (15.4) follows from this and (15.11). The loss which
presumably prevents us from obtaining Halberstam's conjecture
(15.9) arises from having ignored the fact that the sum in
(15.12) should be cancelling: we expect that the bound (15.13)
is weak, at least usually.

We now deduce Theorem 15.2 from Theorem 15.1. The trivial
upper bound (15.8) enables us to say that

$$q E^*(x, q)^2 \ll E^*(x, q) x \log x$$

for $q \leq x$. Hence (15.6) follows immediately from (15.4).
This is well-known, as is the fact that a weaker form of
(15.4) can be deduced from (15.6); from Cauchy's inequality
we have

$$\sum_{q \leq Q} E^*(x, q) \leq \left(\sum_{q \leq Q} q^{-1} \right)^{\frac{1}{2}} \left(\sum_{q \leq Q} q E^*(x, q)^2 \right)^{\frac{1}{2}}.$$

This, with (15.6), gives

$$\sum_{q \leq Q} E^*(x,q) \ll x^{\frac{3}{4}} Q^{\frac{1}{2}} (\log x)^{13},$$

which is a weaker form of (15.4).

We now establish Theorem 15.3. Our main tools are the Siegel - Walfisz theorem (see Davenport, §20) and Theorem 12.2. There is little new in our derivation (see Bombieri [15] and Davenport's book), beyond the fact that Theorem 12.2 permits us to give a slightly stronger result.

If χ is induced by the primitive character χ^* then

$$|\psi(y,\chi) - \psi(y,\chi^*)| \ll (\log q)(\log y) \ll (\log x)^2$$

for $q \leq x$, $y \leq x$, so

$$\sum_{q \leq Q} \phi(q)^{-1} \sum_{\chi \neq \chi_0} \max_{y \leq z} |\psi(y,\chi)|$$

$$\ll Q(\log x)^2 + (\log Q) \sum_{1 < q \leq Q} \phi(q)^{-1} \sum_{\chi}^{*} |\psi(y,\chi)|. \tag{15.14}$$

From the Siegel - Walfisz theorem we see that

$$\sum_{1 < q \leq (\log x)^{3C+42}} \phi(q)^{-1} \sum_{\chi}^{*} \max_{y \leq x} |\psi(y,\chi)| \ll x (\log x)^{-C-1} \tag{15.15}$$

provided $x > x_0(C)$. Hence to prove (15.11) it suffices to show that

$$\sum_{Q \leq q \leq 2Q} \sum_{\chi}^{*} \max_{y \leq x} |\psi(y,\chi)| \ll Q^2 x^{\frac{1}{2}} (\log x)^{11}$$

$$+ Qx (\log x)^{-C-3} \tag{15.16}$$

for $(\log x)^{3C+42} \leq Q \leq x^{\frac{1}{2}}$. From the explicit formula for $\psi(y, \chi)$ we have (see Davenport, p. 163)

$$\psi(y, \chi) \ll \sum_{|\gamma| \leq x^{\frac{1}{2}}} x^{\beta}(1 + |\gamma|)^{-1} + x^{\frac{1}{2}}(\log x)^2. \qquad (15.17)$$

Hence the left hand side of (15.16) is

$$\ll Q^2 x^{\frac{1}{2}}(\log x)^2 + \sum_{Q \leq q \leq 2Q} \sum_{\chi}^{*} (\log x) \int_{\frac{1}{2}-(\log x)^{-1}}^{1} \int_{0}^{2x^{\frac{1}{2}}} N(\sigma, t, \chi)(1 + t^2)^{-1} x^{\sigma} \, dt \, d\sigma.$$

From Theorem 12.2 we see that

$$\sum_{q \leq Q} \sum_{\chi}^{*} N(\sigma, T, \chi) \ll (Q^2 T)^{\frac{5-4\sigma}{3}} (\log QT)^9$$

for $\frac{1}{2} \leq \sigma \leq 1$. Hence the above is

$$\ll Q^2 x^{\frac{1}{2}}(\log x)^2 + (\log x)^{11} \int_{\frac{1}{2}-(\log x)^{-1}}^{1} Q^{\frac{2(5-4\sigma)}{3}} x^{\sigma} \, d\sigma$$

$$\ll Q^2 x^{\frac{1}{2}}(\log x)^2 + (\log x)^{11} (Q^2 x^{\frac{1}{2}} + Q^{\frac{2}{3}} x)$$

$$\ll Q^2 x^{\frac{1}{2}}(\log x)^{11} + Q^{\frac{2}{3}} x (\log x)^{11}.$$

Now if $Q > x^{\frac{2}{5}}$ then the first term is larger, while if $(\log x)^{3C+42} \leq Q \leq x^{\frac{2}{5}}$ then the second term is larger, and it is $\ll Q x (\log x)^{-C-3}$. Hence (15.16) holds, and this with (15.14) and (15.15) gives (15.11).

A lemma in additive prime number theory

In 1937 I.M. Vinogradov [222], [223] (see also [224], Chapter 9) proved that the sum

$$S(\alpha) = S(\alpha, N) = \sum_{n \leqslant N} \Lambda(n) e(n\alpha) \tag{16.1}$$

is "small" when α is not "near" a rational number with small denominator. (Actually Vinogradov treated $\sum_{\mu \leqslant N} e(\mu\alpha)$, which

is equivalent, as one sees by partial summation.) This enabled Vinogradov to show that every large odd number is a sum of three primes. Subsequently van der Corput [36], Čudakov [42], and Estermann [60] showed that Vinogradov's estimates for $S(\alpha)$ enable one to show that almost all positive even integers are sums of two primes. Vinogradov's bound was also essential to Lavrik [108], [109], [110], [111] , [112], [113] in discussing twin primes. In Chapter 17 we use one of Lavrik's results. Linnik [126], [127] and Čudakov [43] (see also [45]) have established bounds similar to Vinogradov's for sums that can be used in place of $S(\alpha)$. Their method depends in a complicated way on bounds for $N(\sigma, T, \chi)$. Here we use theorem 12.1 to give a simple derivation of results which differ insignificantly from those of Vinogradov. We have

THEOREM 16.1. If $S(\alpha)$ is defined by (16.1) and $(a, q) = 1$, then

$$S(\tfrac{a}{q}) \ll \left(q^{\frac{1}{2}}N^{\frac{1}{2}} + N^{\frac{5}{7}}q^{\frac{3}{14}} + Nq^{-\frac{1}{2}}\right)(\log N)^{17}. \tag{16.2}$$

From this we have immediately

COROLLARY 16.2.   If $S(\alpha)$ is defined by (16.1), if
$R \leq q \leq NR^{-1}$, $1 \leq R \leq N^{\frac{1}{4}}$, $(a,q) = 1$, and if
$|\alpha - \tfrac{a}{q}| \leq 2q^{-1}N^{-1}R$, then

$$S(\alpha) \ll NR^{-\frac{1}{2}}(\log N)^{17}. \tag{16.3}$$

Using some simple facts concerning Diophantine
approximations we derive

COROLLARY 16.3.   If $S(\alpha)$ is defined by (16.1), if
$R \leq q \leq NR^{-1}$, $1 \leq R \leq N^{\frac{1}{4}}$, $(a,q) = 1$, and if $|\alpha - \tfrac{a}{q}| \leq q^{-2}$,
then

$$S(\alpha) \ll NR^{-\frac{1}{2}}(\log N)^{17}. \tag{16.4}$$

We now prove the theorem.   We may suppose that $q \leq N$,
for otherwise the estimate is trivial.   We see that

$$S(\tfrac{a}{q}) = \frac{1}{\phi(q)}\sum_{\chi}\Psi(N,\chi)\tau(\overline{\chi})\chi(a) \; + \; O\Big(\sum_{\substack{n \leq N \\ (n,q)>1}}\Lambda(n)\Big)$$

$$\ll q^{-\frac{1}{2}}\log q \sum_{\chi}|\Psi(N,\chi)| \; + \; (\log N)^{2}. \tag{16.5}$$

We see from Theorem 12.1 that

$$\sum_{\chi} N(\sigma, T, \chi) \ll (qT)^{\frac{5-4\sigma}{3}} (\log q T)^{14} \qquad (16.6)$$

for $\frac{1}{2} \le \sigma \le \frac{5}{7}$, and

$$\sum_{\chi} N(\sigma, T, \chi) \ll (qT)^{\frac{5}{2}(1-\sigma)} (\log q T)^{14} \qquad (16.7)$$

for $\frac{5}{7} \le \sigma \le 1$. On the other hand from (15.17) we have

$$\Psi(N, \chi) \ll \sum_{|\gamma| \le N^{\frac{1}{2}}} N^{\beta}(1+|\gamma|)^{-1} + N^{\frac{1}{2}}(\log N)^{2}$$

for $\chi \ne \chi_0$. Hence

$$\sum_{\chi} |\Psi(N, \chi)| \ll N + \sum_{\chi \ne \chi_0} |\Psi(N, \chi)|$$

$$\ll N + q N^{\frac{1}{2}}(\log N)^{2}$$

$$+ \log N \sum_{\chi \ne \chi_0} \int_{\frac{1}{2}-(\log q)^{-1}}^{1} \int_{0}^{2N^{\frac{1}{2}}} N(\sigma, T, \chi)(1+t^{2})^{-1} N^{\sigma} dt d\sigma.$$

This with (16.6) and (16.7) gives

$$\sum_{\chi} |\Psi(N, \chi)| \ll (N^{\frac{1}{2}} q + N^{\frac{5}{7}} q^{\frac{5}{7}} + N)(\log N)^{16},$$

which with (16.5) gives (16.2).

To obtain Corollary 16.2 we sum by parts:

$$S(\alpha) = S(\tfrac{a}{q}) e((\alpha - \tfrac{a}{q})N) - 2\pi i (\alpha - \tfrac{a}{q}) \int_{0}^{N} S(\tfrac{a}{q}, u) e((\alpha - \tfrac{a}{q}) u) du.$$

Hence

$$S(\alpha) \ll |S(\tfrac{a}{q})| + |\alpha - \tfrac{a}{q}| \int_0^N |S(\tfrac{a}{q}, u)| du.$$

But $|\alpha - \tfrac{a}{q}| \leq 2q^{-1}N^{-1}R \leq 2N^{-1}$, so from (16.2) we have

$$S(\alpha) \ll \left(q^{\frac{1}{2}}N^{\frac{1}{2}} + N^{\frac{5}{6}}q^{\frac{2}{14}} + Nq^{-\frac{1}{2}}\right)(\log N)^{17},$$

which is

$$\ll NR^{-\frac{1}{2}}(\log N)^{17}$$

if $R \leq q \leq NR^{-1}$ and $R \leq N^{\frac{1}{4}}$. Hence we have (16.3).

We now prove Corollary 16.3. We assume that $|\alpha - \tfrac{a}{q}| \leq q^{-2}$. If $|\alpha - \tfrac{a}{q}| \leq \tfrac{2R}{qN}$ then we have the desired result immediately from Corollary 16.2. Thus we assume $\tfrac{2R}{qN} \leq |\alpha - \tfrac{a}{q}| \leq q^{-2}$, from which it follows that $q \leq \tfrac{N}{2R}$. By Dirichlet's theorem there is a $q' \leq \tfrac{N}{R}$ such that $|\alpha - \tfrac{a'}{q'}| \leq \tfrac{R}{q'N}$. Hence $\tfrac{1}{qq'} \leq |\tfrac{a}{q} - \tfrac{a'}{q'}| \leq \tfrac{1}{q^2} + \tfrac{R}{q'N}$. As $q \leq \tfrac{N}{2R}$ we have $\tfrac{1}{2qq'} \leq \tfrac{1}{q^2}$, from which it follows that $q' \geq \tfrac{q}{2}$. Hence $\tfrac{R}{2} \leq q' \leq \tfrac{N}{R}$. and Corollary 16.3 now follows from Corollary 16.2.

The mean value theorem of Barban

In Chapter 15 we discussed some results concerning the distribution of prime numbers in arithmetic progressions. We now continue this discussion; we begin by giving a simple proof of the following result, which is slightly sharper than one of Turán [206].

THEOREM 17.1 (Turán).   If the Riemann hypothesis is true of the functions $L(s, \chi)$, $\chi \bmod q$, then

$$\sum_{a=1}^{q} {}^{*} \left(\psi(x; q, a) - \frac{x}{\phi(q)} \right)^{2} \ll x (\log x)^{4}. \tag{17.1}$$

From this we see that if the Riemann hypothesis is true for all $L(s, \chi)$ then

$$\sum_{q \leq Q} q \max_{y \leq x} \sum_{a=1}^{q} {}^{*} \left(\psi(y; q, a) - \frac{y}{\phi(q)} \right)^{2} \ll Q^{2} x (\log x)^{4}. \tag{17.2}$$

As the left hand side is trivially

$$\geq \sum_{q \leq Q} q \left(E^{*}(x, q) \right)^{2},$$

we see that on the GRH one has a result substantially stronger than Theorem 15.2.   The first unconditional approach to (17.2) was made by Barban [5] , [6]   who showed that

$$\sum_{q \leq Q} \sum_{a=1}^{q}{}^{*} \left(\psi(x;q,a) - \frac{x}{\phi(q)} \right)^{2} \ll x^{2} (\log x)^{-A} \qquad (17.3)$$

provided $Q \leq x (\log x)^{-B}$, $B = B(A)$, $x > x_{0}(A)$. Then Davenport and Halberstam [48] gave $B = A + 5$, and Gallagher [68] gave $B = A + 1$. Wilson [227] extended (17.3) to algebraic number fields. All of these results were obtained with the help of the large sieve. We now replace (17.3) by an asymptotic equality, without using the large sieve.

THEOREM 17.2. Let A be positive and $x > x_{0}(A)$. Then for $Q \leq x$

$$\sum_{q \leq Q} \sum_{a=1}^{q}{}^{*} \left(\psi(x;q,a) - \frac{x}{\phi(q)} \right)^{2} = Qx\log x + O\left(Qx \log \frac{2x}{Q} \right)$$
$$+ O\left(x^{2}(\log x)^{-A} \right), \qquad (17.4)$$

and for $Q \geq x$

$$\sum_{q \leq Q} \sum_{a=1}^{q}{}^{*} \left(\psi(x;q,a) - \frac{x}{\phi(q)} \right)^{2} = Qx\log x + \frac{\zeta(2)\zeta(3)}{\zeta(6)} x^{2} \log \frac{Q}{x}$$
$$- Qx + A_{1}x^{2} + O\left(Qx(\log x)^{-A} \right). \qquad (17.5)$$

The first error term in (17.4) may no doubt be made smaller, but is appears that the error is genuinely

$\gg Q x \log\log \frac{x}{Q}$. Halberstam has called my attention to the fact that Barban [6] asserted (17.5) in the case $Q = x$. A.I. Vinogradov informs me that Barban's proof has been criticized, but may yet appear in modified form.

We note that (17.3) is only a power of a logarithm smaller than trivial, like the Bombieri - Vinogradov theorem. For many arithmetic purposes the latter is the appropriate tool, but there are situations in which (17.3) is preferable. Recently Deshouillers (unpublished), used (17.3) to show that if $F(x,y)$ is an irreducible form with integral coefficients, which satisfies some obviously necessary condition, then there are infinitely many pairs of primes p, q for which $F(p,q) = P_{d+1}$ where d is the degree of F .

We now prove Theorem 17.1. We consider first the uninteresting case $q \geq x$. In this case we have

$$\sum_{a=1}^{q} {}^{*} \left(\psi(x;q,a) - \frac{x}{\phi(q)} \right)^2 \ll \frac{x^2}{\phi(q)} + \sum_{n \leq x} \Lambda(n)^2$$

$$\ll \frac{x^2}{q} \log q + x \log x$$

$$\ll x \log x ,$$

so we have (17.1). We now suppose that $q \leq x$. We have

$$\sum_{a=1}^{q} {}^{*} \left(\psi(x;q,a) - \frac{x}{\phi(q)} \right)^2 = \frac{(\psi(x,\chi_0) - x)^2}{\phi(q)} + \frac{1}{\phi(q)} \sum_{\chi \neq \chi_0} |\psi(x,\chi)|^2. \qquad (17.6)$$

Now $\psi(x, \chi_0) = \psi(x) + O((\log q)(\log x))$, and if the Riemann
hypothesis holds then $\psi(x) - x \ll x^{\frac{1}{2}}(\log x)^2$, so
$\psi(x, \chi_0) - x \ll x^{\frac{1}{2}}(\log x)^2$. Similarly, if $\chi \neq \chi_0$ and χ^*
is the primitive character which induces χ then
$\psi(x, \chi) = \psi(x, \chi^*) + O((\log q)(\log x))$, and if $L(s, \chi)$ has
no zeros in $\sigma > \frac{1}{2}$ then $\psi(x, \chi^*) \ll x^{\frac{1}{2}}(\log q x)^2$, so
$\psi(x, \chi) \ll x^{\frac{1}{2}}(\log x)^2$. These estimates, in (17.6), imply
(17.1).

We now prove Theorem 17.2 (as in [144]).

Although we derive Theorem 17.2 without appealing to the
large sieve, we require a result of Lavrik [109] (see also
[109] , [110] , [111] , [112], [113]) concerning twin primes on
average. We let

$$\mathfrak{S} = 2 \prod_{p > 2} (1 - \frac{1}{(p-1)^2}), \tag{17.7}$$

for $k > 0$ we let

$$\psi_2(x, k) = \sum_{k < n \le x} \Lambda(n) \Lambda(n-k), \tag{17.8}$$

and finally we set

$$E(x, k) = \psi_2(x, 2k) - \mathfrak{S}(x - 2k) \prod_{\substack{p \mid k \\ p > 2}} \left(\frac{p-1}{p-2}\right). \tag{17.9}$$

Lavrik's result, after partial summation, is

LEMMA 17.3 (Lavrik). Let $\mathfrak{S}$, $\psi_2(x, k)$ and $E(x, k)$
be defined by (17.7), (17.8), and (17.9). Then for $B > 0$
and $x > x_0(B)$

$$\sum_{k=1}^{\frac{x}{2}} (E(x,k))^2 \ll x^2 (\log x)^{-B}. \tag{17.10}$$

We also require

LEMMA 17.4.   Let $\mathfrak{S}$ be defined as in (17.7).   For $y \geqslant 2$ we have

$$\sum_{\substack{m \leq y}} \prod_{\substack{p \mid m \\ (p, 2r) = 1}} \left(\frac{p-1}{p-2} \right) = \frac{2y}{\mathfrak{S}} \prod_{\substack{p \mid r \\ p > 2}} \left(1 - \frac{1}{(p-1)^2} \right) + O(\log y), \tag{17.11}$$

and

$$\sum_{\substack{m \leq y}} m \prod_{\substack{p \mid m \\ (p, 2r) = 1}} \left(\frac{p-1}{p-2} \right) = \frac{y^2}{\mathfrak{S}} \prod_{\substack{p \mid r \\ p > 2}} \left(1 - \frac{1}{(p-1)^2} \right) + O(y \log y), \tag{17.12}$$

uniformly for positive integers r.

Our proof of these estimates follows in spirit the elementary proof that

$$\sum_{n \leq x} \phi(n) = \frac{3}{\pi^2} x^2 + O(x \log x), \tag{17.13}$$

though we must ensure that our error terms are uniform. It is interesting that while the error term in (17.13) may be replaced([226], p. 114) by $O(x (\log x)^{\frac{2}{3}} (\log\log x)^{\frac{4}{3}})$, deeper arguments show that the error term in (17.11) cannot be made smaller.

Finally we require

LEMMA 17.5. <u>For</u> $x^{\frac{1}{2}} \leq Q \leq x$

$$\sum_{q \leq Q} \sum_{\substack{k \leq \frac{x}{q} \\ 2 \mid qk}} (x - qk) \prod_{\substack{p \mid qk \\ p > 2}} \left(\frac{p-1}{p-2}\right) = \frac{x^2}{2\mathfrak{S}} \sum_{q \leq Q} \frac{1}{\phi(q)} + O\left(Qx \log \frac{2x}{Q}\right), \qquad (17.14)$$

<u>and for</u> $Q \geq x$

$$\sum_{q \leq Q} \sum_{\substack{k \leq \frac{x}{q} \\ 2 \mid qk}} (x - qk) \prod_{\substack{p \mid qk \\ p > 2}} \left(\frac{p-1}{p-2}\right) = \frac{\mathfrak{I}(2)\mathfrak{I}(3)}{\mathfrak{I}(6)} x^2 \log x + A_2 x^2$$

$$+ O\left(x^{\frac{3}{2}} \log x\right). \qquad (17.15)$$

To prove Theorem 17.2 we use Lemmas 17.3, 17.4 and 17.5. We do not include a proof of Lemma 17.3, though the results of Chapter 16 would make this a simple matter. We first prove Theorem 17.2, and then derive (the less interesting) Lemmas 17.4 and 17.5.

We note that the inner sum on the left hand side of (17.4) is

$$= \sum_{a=1}^{q} {}^{*} \left(\psi(x; q, a)\right)^2 + \frac{x}{\phi(q)} \left(x - 2 \sum_{\substack{n \leq x \\ (n, q) = 1}} \Lambda(n)\right)$$

$$= \sum_{n_1 \leq x} \sum_{\substack{n_2 \leq x \\ n_1 \equiv n_2 (\bmod q) \\ (n_1, n_2, q) = 1}} \Lambda(n_1)\Lambda(n_2) - \frac{x^2}{\phi(q)}$$

$$+ O\left(\frac{x^2}{\phi(q)(\log x)^5}\right) + O\left(\frac{x(\log x)^2}{\phi(q)}\right),$$

by the prime number theorem. If in the double sum we drop

the condition $(n, n_2, q) = 1$ then we increase it by an amount which is $\ll (\log q)^3$. Thus the left hand side of (17.4) or (17.5) is

$$= Q \sum_{n \leq x} (\Lambda(n))^2 + 2 \sum_{q \leq Q} \sum_{\substack{n_1 < n_2 \leq x \\ n_1 \equiv n_2 (q)}} \Lambda(n_1) \Lambda(n_2) \quad - x^2 \sum_{q \leq Q} \frac{1}{\phi(q)}$$

$$+ O\left(\frac{x^2 \log Q}{(\log x)^B}\right) + O\left(x (\log Q)^3\right) + O\left(Q(\log Q)^3\right). \tag{17.16}$$

Now

$$\sum_{n \leq x} (\Lambda(n))^2 = x \log x - x + O\left(x (\log x)^{-A}\right) \tag{17.17}$$

by the prime number theorem. In addition,

$$2 \sum_{q \leq Q} \sum_{\substack{n_1 < n_2 \leq x \\ n_1 \equiv n_2 (\bmod q)}} \Lambda(n_1) \Lambda(n_2) = 2 \sum_{q \leq Q} \sum_{k \leq \frac{x}{q}} \psi_2(x; q k).$$

If r is odd then $\psi_2(x, r) \ll (\log x)^2$; therefore odd values of r contribute no more than $\ll x (\log x)^3$. Hence from (17.9) the above is

$$= 2 \mathfrak{S} \sum_{q \leq Q} \sum_{\substack{q k \leq x \\ 2 | q k}} (x - q k) \prod_{\substack{p | q k \\ p > 2}} \left(\frac{p-1}{p-2}\right) + O\left(\sum_{2m \leq x} d(2m) |E(x, m)|\right)$$

$$+ O\left(x (\log x)^3\right). \tag{17.18}$$

By Cauchy's inequality the first error term is

$$\ll \left(\sum_{2m \leq x} d^2(m)\right)^{\frac{1}{2}} \left(\sum_{2m \leq x} E(x, m)^2\right)^{\frac{1}{2}} .$$

The first factor is $\ll x^{\frac{1}{2}}(\log x)^{\frac{3}{2}}$, and the second factor is $\ll x^{\frac{1}{2}}(\log x)^{-\frac{B}{2}}$, by Lemma 17.3. We take $B = 2A+3$, so that this error is $\ll x(\log x)^{-B}$. To prove (17.4) we have only to combine (17.16), (17.17), and (17.18) with (17.14). To obtain (17.5) we combine (17.16), (17.17), (17.18), and (17.15) with the fact that

$$\sum_{m \leq M} \frac{1}{\phi(m)} = \frac{\zeta(2)\zeta(3)}{\zeta(6)} \log M + A_3 + O\left(\frac{\log M}{M}\right).$$

We now prove (17.11) of Lemma 17.4. Let

$$f_r(m) = \begin{cases} \mu^2(m) \prod_{p|m} (p-2)^{-1} & \text{if } (m, 2r) = 1 \\ \\ 0 & \text{if } (m, 2r) > 1. \end{cases}$$

Then the left hand side of (17.11) is

$$= \sum_{m \leq y} \sum_{d|m} f_r(d) = \sum_{d \leq y} f_r(d) \sum_{\substack{m \leq y \\ d|m}} 1$$

$$= y \sum_{d \leq y} \frac{f_r(d)}{d} + O\left(\sum_{d \leq y} f_r(d)\right)$$

$$= y \sum_{d=1}^{\infty} \frac{f_r(d)}{d} + O\left(y \sum_{d > y} \frac{f_r(d)}{d}\right) + O\left(\sum_{d \leq y} f_r(d)\right)$$

$$= \frac{2y}{\mathfrak{S}} \prod_{\substack{p|r \\ p > 2}} \left(1 - \frac{1}{(p-1)^2}\right) + O(\log y).$$

This is (17.11). We obtain (17.12) from (17.11) by partial

summation.

We now prove Lemma 17.5. On the left hand side of (17.14) and (17.15) the condition $2|qk$ may be expressed as $2|q$ or $2|k$. Therefore the expression is

$$= 2F(Q, \tfrac{x}{2}) + 2F(\tfrac{Q}{2}, \tfrac{x}{2}) - 4F(\tfrac{Q}{2}, \tfrac{x}{4}), \tag{17.19}$$

where

$$F(U, y) = \sum_{u \leq U} \sum_{v = \frac{y}{u}} (y - uv) \prod_{\substack{p|uv \\ p>2}} \left(\frac{p-1}{p-2}\right).$$

Now for $U \leq y$

$$F(U, y) = \sum_{u \leq U} \prod_{\substack{p|u \\ p>2}} \left(\frac{p-1}{p-2}\right) \sum_{v \leq \frac{y}{u}} (y - uv) \prod_{\substack{p|v \\ (p, 2u)=1}} \left(\frac{p-1}{p-2}\right);$$

we apply Lemma 17.3 and simplify, to find that this is

$$= \frac{y^2}{6} \sum_{u \leq U} u^{-1} \prod_{\substack{p|u \\ p>2}} \frac{p}{p-1} + O\left(y \sum_{u \leq U} (\log \tfrac{y}{u}) \prod_{\substack{p|u \\ p>2}} \left(\frac{p-1}{p-2}\right)\right)$$

$$= \frac{y^2}{6} \sum_{u \leq U} \frac{1}{\phi(2u)} + O\left(y \, U \log \tfrac{3y}{U}\right).$$

We can now weaken the condition that $U \leq y$ to $U \leq 2y$, as the difference may be absorbed into the error term. Taking this result with (17.19), we have (17.14). More careful treatments of $F(U, y)$ can be made along the lines of the elementary treatment of

$$\sum_{n \leq x} d(n),$$

and all the error terms are small, except in estimating

$$y^2 \sum_{u \leq \frac{x}{U}} \frac{1}{\phi(2u)} \; .$$

One cannot hope to estimate this with an error less than $O(y \, U \log \log \frac{x}{U})$. Hence the error term in (17.14) cannot be $\ll Q x$ if $Q = o(x)$.

If $U \geq y$ then $F(U,y)$ does not depend on U , and in fact

$$F(U,y) = 2 \sum_{u \leq y^{\frac{1}{2}}} \prod_{\substack{p \mid u \\ p > 2}} \left(\frac{p-1}{p-2} \right) \sum_{v \leq \frac{y}{u}} (y - uv) \prod_{\substack{p \mid v \\ (p, 2u) = 1}} \left(\frac{p-1}{p-2} \right)$$

$$- \sum_{u \leq y^{\frac{1}{2}}} \prod_{\substack{p \mid u \\ p > 2}} \left(\frac{p-1}{p-2} \right) \sum_{v \leq y^{\frac{1}{2}}} (y - uv) \prod_{\substack{p \mid v \\ (p, 2u) = 1}} \left(\frac{p-1}{p-2} \right) ,$$

and by an appeal to Lemma 17.3 we find that this is

$$= \frac{2 y^2}{\mathfrak{S}} \sum_{u \leq y^{\frac{1}{2}}} \frac{1}{\phi(2u)} - \frac{2 y^{\frac{3}{2}}}{\mathfrak{S}} \sum_{u \leq y^{\frac{1}{2}}} \prod_{\substack{p \mid u \\ p > 2}} \frac{p}{p-1}$$

$$+ \frac{y}{\mathfrak{S}} \sum_{u \leq y^{\frac{1}{2}}} u \prod_{\substack{p \mid u \\ p > 2}} \frac{p}{p-1} + O(y^{\frac{3}{2}} \log y)$$

By results similar to those of Lemma 17.3, this is

$$= \frac{y^2 \log y}{\mathfrak{S}} \prod_{p > 2} \left(1 + \frac{1}{p(p-1)} \right) + A_4 y^2 + O(y^{\frac{3}{2}} \log y).$$

This, with (17.19), gives (17.15).

A P P E N D I X   I

We prove the inequality

$$(\sin \pi x)^{-2} \le (\pi \|x\|)^{-2} + 1,\qquad\qquad(\text{I.1})$$

which is required in the proofs of Theorems 2.3 and 2.4. This inequality is essentially trivial; we include its proof only because (I.1) is slightly stronger (and tidier) than one might expect.

To prove (I.1) it suffices to show that

$$(\sin x)^{-2} \le x^{-1} + 1$$

for $0 < x \le \frac{\pi}{2}$.   We have

$$\sin x = x - \frac{x^3}{3!} + \frac{x^5}{5!} - \dots .$$

If $\frac{x^2}{6.7} \le 1$ then the terms are of diminishing magnitude from the x^5 term on, so

$$\sin x > x - \frac{x^3}{6}$$

for $0 < x \le \frac{\pi}{2}$.   Thus

$$(\sin x)^{-1} < x^{-1}\left(1 - \frac{x^2}{6}\right)^{-1}$$

$$= x^{-1}\left(1 + \frac{x^2}{6}\left(1 - \frac{x^2}{6}\right)^{-1}\right).$$

As $x^2 \le \left(\frac{\pi}{2}\right)^2 < \frac{5}{2}$ we have $\left(1 - \frac{x^2}{6}\right)^{-1} < \frac{12}{7}$, so our bound is

$$< x^{-1}\left(1 + \frac{2}{7}x^2\right).$$

Hence

$$(\sin x)^{-2} \leq x^{-2}\left(1 + x^2\left(\tfrac{4}{7} + \tfrac{4}{49}x^2\right)\right)$$

$$< x^{-2}\left(1 + x^2\left(\tfrac{4}{7} + \tfrac{4}{49}\cdot\tfrac{5}{2}\right)\right)$$

$$< x^{-2} + 1.$$

Alternatively, we can proceed from the identity

$$(\sin x)^{-2} = \sum_{n=-\infty}^{+\infty} (x - \pi n)^{-2}.$$

From the above we have

$$(\sin \pi x)^{-2} - (\pi x)^{-2} = \frac{2}{\pi^2}\sum_{n=1}^{\infty} \frac{n^2 + x^2}{(n^2 - x^2)^2}.$$

Here the right hand side is obviously increasing for $0 \leq x \leq \tfrac{1}{2}$. Hence in this range

$$(\sin \pi x)^{-2} - (\pi x)^{-2} \leq \left(\sin \tfrac{\pi}{2}\right)^{-2} - \left(\tfrac{\pi}{2}\right)^{-2}$$

$$= 1 - \tfrac{4}{\pi^2}.$$

Thus in general

$$(\sin \pi x)^{-2} \leq (\pi \|x\|)^{-2} + 1 - \tfrac{4}{\pi^2}.$$

We prove here two results for use in Chapter 8.

THEOREM II.1.   Let χ be a character mod q, and let $s = \sigma + it$ satisfy $\sigma \geqslant 0$.   We put $\tau = |t| + 2$.   Then

$$\sum_{n \leqslant N} \left(1 - \frac{n}{N}\right) \chi(n) n^{-s} \ll (q\tau)^{\frac{1}{2}} \log q\tau + \varepsilon(\chi) N \tau^{-2}, \qquad (II.1)$$

where $\varepsilon(\chi) = 1$ or 0 according as χ is principal or not.

We note that when $s = 0$ the bound (II.1) is (averaging aside) the inequality of Pólya [151] and Vinogradov [221]. Pólya's proof depends on an identity which arises from the consideration of the sum as a periodic function.   Later Knapowski (unpublished) gave a second derivation of Pólya's identity. Knapowski's approach was similar to the one which has been employed by Chandrasekharan and Narasimhan [33] and Lavrik [115] , [119] , for proving approximate functional equations.   We proceed along the same lines, but we avoid the need of establishing an identity.   We use (II.1) in proving Theorems 8.2, 8.3, and we use the following result in proving Theorem 8.4.

THEOREM II.2.   If $s = \sigma + it$ satisfies $\sigma \geqslant 0$ then

$$\sum_{n=1}^{\infty} \left(e^{-\frac{n}{2N}} - e^{-\frac{n}{N}}\right) n^{-s} \ll N^{\theta} M(\theta, |2t|) + N e^{-|t|} \qquad (II.2)$$

for $0 \leqslant \theta \leqslant 1$,   where $M(\alpha, T)$ is defined by (8.9).

The following lemma saves us work by enabling us to appeal
to known facts about the zeta function.

LEMMA II.3.   If

$$f(s) = \sum_{n=1}^{\infty} a_n n^{-s}$$

and

$$g(s) = \sum_{n=1}^{\infty} b_n n^{-s}$$

are absolutely convergent when $\sigma = \sigma_0$, and if $|a_n| \le b_n$
for all n, then

$$\int_{-T}^{T} |f(\sigma_0+it)|^2 dt \le 2\int_{-2T}^{2T} |g(\sigma_0+it)|^2 dt$$

for any $T \ge 0$.

The proof is very simple.   We have

$$\int_{-T}^{T} |f(\sigma_0+it)|^2 dt \le 2\int_{-2T}^{2T} \left(1-\frac{|t|}{2T}\right)|f(\sigma_0+it)|^2 dt$$

$$= 2\sum_{m=1}^{\infty}\sum_{n=1}^{\infty} \frac{a_m \overline{a_n}}{(mn)^{\sigma_0}} \int_{-2T}^{2T}\left(1-\frac{|t|}{2T}\right)\left(\frac{m}{n}\right)^{it} dt$$

$$\le 2\sum_{m=1}^{\infty}\sum_{n=1}^{\infty} \frac{b_m b_n}{(mn)^{\sigma_0}} \int_{-2T}^{2T}\left(1-\frac{|t|}{2T}\right)\left(\frac{m}{n}\right)^{-it} dt$$

$$= 2\int_{-2T}^{2T} \left(1-\frac{|t|}{2T}\right)|g(\sigma_0+it)|^2 dt$$

$$\le 2\int_{-2T}^{2T} |g(\sigma_0+it)|^2 dt,$$

159

since $\int_{-1}^{1}(1-|t|)y^{it}dt \geq 0$ for any real y .

We now prove Theorem II.1. Summing by parts twice, we see that it suffices to show that (II.1) holds when Now

$$\sum_{n\leq N}\left(1-\frac{n}{N}\right)\chi(n)n^{-it} = \frac{1}{2\pi i}\int_{2-i\infty}^{2+i\infty} L(w+it,\chi)N^{w}w^{-1}(w+1)^{-1}dw,$$

and we take the contour to the abscissa $a = -(\log q\tau)^{-1}$. In doing this we pass a simple pole at $w=0$, and also one at $w=1-it$ if χ is principal. The residues are $\ll |L(it,\chi)|$ and $\ll N\tau^{-2}$ respectively. We suppose that χ is induced by χ_1 mod q_1 , and we write $q_1 q_2 = q$. From the functional equation for $L(s,\chi_1)$ (see Davenport, §.9) we see that

$$L(it,\chi) \ll (q_1\tau)^{\frac{1}{2}}d(q_2)|L(1-it,\bar{\chi}_1)|$$

$$\ll (q_1\tau)^{\frac{1}{2}}d(q_2)\log q_1\tau$$

$$\ll (q\tau)^{\frac{1}{2}}\log q\tau,$$

and that for w on our new contour

$$L(w+it,\chi) \ll (q\tau)^{\frac{1}{2}}d(q_2)|L(1-w,\bar{\chi}_1)| \ll (q\tau)^{\frac{1}{2}}|L(1-w-it,\bar{\chi}_1)|.$$

Hence $\sum_{n\leq N}\left(1-\frac{n}{N}\right)\chi(n)n^{it} \ll (q\tau)^{\frac{1}{2}}\log q\tau + \varepsilon(\chi)N\tau^{-2}$

$$+ (q\tau)^{\frac{1}{2}}\int_{a-i\infty}^{a+i\infty}|L(1-w-it,\bar{\chi}_1)|w^{-1}(w+1)^{-1}||dw|.$$

By the Cauchy - Schwarz inequality

$$\int |L(1-w-it,\overline{\chi}_1)\, w^{-1}(w+1)^{-1}| \,|dw| \leq \left(\int \frac{|w|^2+1}{|w(w+1)|^2}|dw|\right)^{\frac{1}{2}} \left(\int \frac{|L(1-w-it,\overline{\chi}_1)|^2}{1+|w|^2}|dw|\right)^{\frac{1}{2}}. \quad (II.3)$$

Now by Lemma II.3

$$\int_{a+in}^{a+i(n+1)} |L(1-w-it,\overline{\chi}_1)|^2 |dw| \leq 2 \int_{1-a-i}^{1-a+i} |\mathfrak{Z}(w)|^2 |dw|$$

$$\ll \int_{1-a-i}^{1-a+i} \frac{|dw|}{|w-1|^2}$$

$$\ll \log q\tau,$$

so the integral in the second factor on the right hand side of (II.3) is $\ll \log q\tau$. The integral in the first factor on the right is also $\ll \log q\tau$, so we have (II.1).

To prove Theorem II.2 we note that as above it suffices to consider the case $\sigma = 0$. We write

$$\sum_{n=1}^{\infty} \left(e^{-\frac{n}{2N}}-e^{-\frac{n}{N}}\right) n^{-it} = \frac{1}{2\pi i} \int_{2-i\infty}^{2+i\infty} \mathfrak{Z}(w+it)\,\Gamma(w)\left((2N)^w-N^w\right)dw.$$

We take the contour to the abscissa θ, making a detour around $w = 1-it$ if θ is near 1 . We have a residue of $N\,\Gamma(1-it) \ll N\,e^{-|t|}$ from the pole in the integrand at $w = 1-it$. For $|\operatorname{dm}w| \leq t$ the integrand is $\ll N^\theta e^{-|\operatorname{dm}w|}\, M(\theta,2|t|)$, while otherwise it is $\ll N^\theta e^{-|\operatorname{dm}w|} \ll N^\theta M(\theta,2|t|)\, e^{-|\operatorname{dm}w|}$. Hence

$$\sum_{n=1}^{\infty} \left(e^{-\frac{n}{2N}} - e^{-\frac{n}{N}} \right) n^{-it} \ll N e^{-|t|} + N^{\theta} M(\theta, 2|t|) \int_{-\infty}^{+\infty} e^{-|u|} \, du,$$

and the result follows.

B I B L I O G R A P H Y

1 Ankeny, N.C., "The least quadratic non residue," Ann. of Math., 55 (1952), 65-72.

2 Atkinson, F.V., "The mean value of the zeta-function on the critical line," Proc. London Math. Soc., (2) 47 (1941), 174-200.

3 Atkinson, F.V., "The mean value of the Riemann zeta function," Acta Math., 81 (1949), 353-376.

4 Barban, M.B., "The density of zeros of Dirichlet L-series and the problem of sums of prime and almost prime numbers," Mat. Sb., 61 (103) (1963), 418-425.

5 Barban, M.B., "Analogues of the divisor problem of Titchmarsh," Vestnik Leningrad. Univ. Ser. Mat. Meh. Astronom., 18 (1963), no. 4, 5-13.

6 Barban, M.B., "On the average error in the generalized prime number theorem," Dokl. Akad. Nauk UzSSR, (1964), No. 5, 5-7.

7 Barban, M.B., "The 'large sieve' method and its appli- cations in the theory of numbers," Uspehi Mat. Nauk, 21 (1966), 51-102. See also Russian Math. Surveys, 21 (1966), no. 1, 49-103.

8 Bateman, P.T., S. Chowla and P. Erdős, "Remarks on the size of $L(1,\chi)$," Publ. Math. Debrecen, 1 (1949- 1950), 165-182.

9 Bellman, Richard, "Almost orthogonal series," Bull. Amer. Math. Soc., (2) 50 (1944), 517-519.

10 Bellman, Richard, "Wigert's approximate functional equation and the Riemann zeta function," Duke Math. J., 16 (1949), 547-552.

11 Boas, R.P., "A general moment problem," Amer. J. Math., 63 (1941), 361-370.

12 Bohr, H., and E. Landau, "Beiträge zur Theorie der Riemannschen Zetafunktion," Math. Ann., 74 (1913), 3-30.

13 Bohr, H., and E. Landau, "Ein Satz über Dirichletsche Reihen mit Anwendung auf die ʃ-Funktion und die L-Funktionen," Rend. Circ. Mat. Palermo, 37 (1914), 269-272.

14 Bohr, H., and E. Landau, "Sur les zeros de la fonction
    ζ(s) de Riemann," C. R. Acad. Sci. Paris, 158
    (1914), 106-110.

15 Bombieri, E., "On the large sieve," Mathematika, 12
    (1965), 201-225.

16 Bombieri, E., "A remark on the large sieve," Proc. Rome
    Number Theory Conference, 1968.

17 Bombieri, Enrico, "Density theorems for the zeta function,"
    Proceedings of the Stony Brook Number Theory
    Conference, 1969, American Mathematical Society,
    Providence, to appear.

18 Bombieri, E., "A note on the large sieve," Acta Arith.,
    to appear.

19 Bombieri, E., and H. Davenport, "Small differences
    between prime numbers," Proc. Roy. Soc. Ser. A,
    293 (1966), 1-18.

20 Bombieri, E., and H. Davenport, "On the large sieve
    method," Abh. aus Zahlentheorie und Analysis
    Zur Erinnerung an Edmund Landau, Deut. Verlag
    Wiss., Berlin, 1968, 11-22.

21 Bombieri, E., and H. Davenport, "Some inequalities
    involving trigonometrical polynomials," Ann.
    Scuola Norm. Sup. Pisa, 23 (1969), 223-241.

22 Brauer, A., "Limits for the characteristic roots of a
    matrix, I, II, IV," Duke Math. J., 13 (1946),
    387-395; 14 (1947), 21-26; 19 (1952), 73-91.

23 Browne, E.T., "The characteristic roots of a matrix,"
    Bull. Amer. Math. Soc., 36 (1930), 705-710.

24 Burgess, D.A., "The distribution of quadratic residues
    and non-residues," Mathematika, 4 (1957),
    106-112.

25 Burgess, D.A., "On character sums and primitive roots,"
    Proc. London Math. Soc., (3) 12 (1962),
    179-192.

26 Burgess, D.A., "On character sums and L-series, I, II,"
    Proc. London Math. Soc., (3) 12 (1962),
    193-206; (3) 13 (1963), 524-536.

27 Burgess, D.A., "A note on the distribution of residues
    and non-residues,"   J. London Math. Soc.,
    38 (1963), 253-256.

28 Burgess, D.A., "A note on L-functions,"   J. London
      Math. Soc., 39 (1964), 103-108.

29 Burgess, D.A., "A form of the large sieve," Proc.
      Oberwohlfach Number Theory Conference, 1970,
      to appear.

30 Burgess, D.A., and P.D.T.A. Elliott, "The average of
      the least primitive root," Mathematika, 15
      (1968), 39-50.

31 Carlson, F., "Über die Nullstellen der Dirichletschen
      Reihen und der Riemannschen ζ-funktionen,"
      Arkiv för Mat. Astronom. och Fysik, 15 (1920),
      No. 20.

32 Cassels, J.W.S., "Footnote to a note of Davenport and
      Heilbronn," J. London Math. Soc., 36 (1961),
      177-184.

33 Chandrasekharan, K., and Raghavan Narasimhan, "The
      approximate functional equation for a class of
      zeta-functions," Math. Ann., 152 (1963), 30-64.

34 Choi, S.L.G., "On a theorem of Roth," Math. Ann., 179
      319-328.

35 Cohen, Eckford, "An extension of Ramanujan's sum,"
      Duke Math. J., 16 (1949), 85-90.

36 Corput, J.G. van der, "Sur l'hypothèse de Goldbach,"
      Proc. Akad. Wet. Amsterdam, 41 (1938), 76-80.

37 Cramér, H., "Some theorems concerning prime numbers,"
      Arkiv för Mat. Astronom. och Fysik, 15 (1920),
      no. 5, 1-32.

38 Cramér, H., "On the distribution of primes," Proc.
      Cambridge Philos. Soc., 20 (1921), 272-280.

39 Cramér, H., "Ein Mittelwertsatz in der Primzahltheorie,"
      Math. Z., 12 (1922), 147-153.

40 Cramér, Harald, "On the order of magnitude of the
      difference between consecutive prime numbers,"
      Acta Arith., 2 (1937), 23-46.

41 Čudakov, N., "On the difference between two neighboring
      prime numbers," Mat. Sb., 1 (1936), 799-814.

42 Čudakov, N., "On Goldbach's problem," Dokl. Akad. Nauk
      SSSR, 17 (1937), 331-334.

43 Čudakov, N., "On Goldbach-Vinogradov's theorem," Ann.
        Math., 48 (1947), 515-545.

44 Čudakov, N.G., "On the limits of variation of the
        function $\psi(x, k, 1)$," Izv. Akad. Nauk SSSR Ser.
        Mat., 12 (1948), 31-46.

45 Čudakov, N.G., and K.A. Rodosskiĭ, "New methods in the
        theory of Dirichlet's L-functions," Uspehi
        Mat. Nauk (N. S.), 4 (1949), no. 2(30), 22-56.
        See also Amer. Math. Soc. Transl., no. 73,
        1952, 44 pp.

46 Davenport, Harold, Multiplicative number theory, Markham,
        Chicago, 1967.   (Delete Theorems 4, 4A in §23.)

47 Davenport, H., and H. Halberstam, "The values of a
        trigonometric polynomial at well spaced points,"
        Mathematika, 13 (1966), 91-96.   See also:
        "Corrigendum and addendum," Mathematika, 14
        (1967), 229-232.

48 Davenport, H., and H. Halberstam, "Primes in arithmetic
        progressions," Michigan Math. J., 13 (1966),
        485-489.   See also: "Corrigendum," Michigan
        Math. J., 15 (1968), 505.

49 Davenport, H., and H. Heilbronn, "On the zeros of certain
        Dirichlet series, I, II," J. London Math. Soc.,
        11 (1936), 181-185, 307-312.

50 Elliott, P.D.T.A., "On the size of $L(1, \chi)$," J. Reine
        Angew. Math., 236 (1969), 26-36.

51 Elliott, P.D.T.A., "A restricted mean value theorem,"
        J. London Math. Soc., (2) 1 (1969), 447-460.

52 Elliott, P.D.T.A., "A conjecture of Erdős concerning
        character sums," Nederl. Akad. Wetensch. Proc.
        Ser. A, 72 (1969), (= Indag. Math., 31 (1969)),
        164-171.

53 Elliott, P.D.T.A., "The distribution of primitive roots,"
        Canad. J. Math., 21 (1969), 822-841.

54 Elliott, P.D.T.A., "Some applications of a theorem of
        Raikov to number theory," J. Number Theory, 2
        (1970), 22-55.

55 Elliott, P.D.T.A., "The Turán-Kubilius inequality, and
        a limitation theorem for the large sieve," Amer.
        J. Math., 92 (1970), 293-300.

56 Elliott, P.D.T.A., "On the mean value of f(p)," Proc. London Math. Soc., (3) 21 (1970), 28-96.

57 Elliott, P.D.T.A., and Halberstam, H., "Some applications of Bombieri's theorem," Mathematika, 13 (1966), 196-203.

58 Erdős, P., "Remarks on number theory, V.," Mat. Lapok, 17 (1966), 135-155.

59 Erdős, P., and A. Rényi, "Some remarks on the large sieve of Yu. V. Linnik," Ann. Univ. Sci. Budapest., 11 (1968), 3-13.

60 Estermann, T., "On Goldbach's problem: Proof that almost all even positive integers are sums of two primes," Proc. London Math. Soc., (2) 44 (1938), 307-314.

61 Estermann, T., Introduction to modern prime number theory, Cambridge Mathematical Tracts No. 41, Cambridge, 1961.

62 Farnell, A.B., "The characteristic roots of a matrix," Bull. Amer. Math. Soc., 50 (1944), 789-794.

63 Fogels, E., "On average values of arithmetical functions," Acta. Univ. Latviensis, 3 (1940), 285-313 (= Publ. Sem. Math. Univ. Lettonie, no. 16, 1940).

64 Fogels, E., "On average values of arithmetic functions," Proc. Cambridge Philos. Soc., 37 (1941), 358-372.

65 Fogels, E., "On the zeros of L-functions," Acta Arith., 11 (1965), 67-96.

66 Fridlender, V.R., "On the least nth-power non-residue," Dokl. Akad. Nauk SSSR 66 (1949), 351-352.

67 Gabriel, R.M., "Some results concerning the integrals of moduli of regular functions along certain curves," J. London Math. Soc., 2 (1927), 112-117.

68 Gallagher, P.X., "The large sieve," Mathematika, 14 (1967), 14-20.

69 Gallagher, P.X., "Bombieri's mean value theorem," Mathematika, 15 (1968), 1-6.

70 Gallagher, P.X., "A larger sieve," Proceedings of the Stony Brook Number Theory Conference, 1969, American Mathematical Society, Providence, to appear.

71 Gallagher, P.X., "A large sieve density estimate near
    $\sigma = 1$," Invent. Math., to appear.

72 Golomb, Solomon W., "The lambda method in prime number
    theory," J. Number Theory, 2 (1970), 193-198.

73 Halász, G., "Über die Mittelwerte multiplikativer
    zahlentheoretischer Funktionen," Acta Math.
    Acad. Sci. Hungar., 19 (1968), 365-403.

74 Halász, G., "On the average order of magnitude of
    Dirichlet series," Acta Math. Acad. Sci. Hungar.,
    21 (1970), 227-233.

75 Halász, G., and P. Turán, "On the distribution of roots
    of Riemann zeta and allied functions, I," J.
    Number Theory, 1 (1969), 121-137.

76 Halberstam, H., W. Jurkat, and H. -E. Richert, "Un
    nouveau resultat de la methode du crible,"
    C. R. Acad. Sci. Paris Ser. A, 264 (1967),
    920-923.

77 Halberstam, H., and K.F. Roth, Sequences, vol I, Oxford:
    Clarendon Press, 1966.

78 Halmos, Paul R., Finite-dimensional vector spaces, Second
    Ed., D. Van Nostrand, New York, 1958.

79 Haneke, W., "Verschärfung der Abschätzung von $\zeta(\frac{1}{2} + it)$,"
    Acta Arith., 8 (1963), 357-430.

80 Hardy, G.H., and J.E. Littlewood, "Contributions to the
    theory of the Riemann zeta-function and the theory
    of the distribution of primes," Acta Math., 41
    (1918), 119-196.

81 Hardy, G.H., J.E. Littlewood, and G. Pólya, Inequalities,
    Second Ed., Cambridge University Press, 1964.

82 Hardy, G.H., and E.M. Wright, An introduction to the
    theory of numbers, Fourth Ed., Oxford Clarendon
    Press, 1964.

83 Haselgrove, C.B., "Some theorems in the analytic theory
    of numbers," J. London Math. Soc., 26 (1951),
    273-277.

84 Heilbronn, H., "Über den Primzahlsatz von Herrn Hoheisel,"
    Math. Z., 36 (1933), 394-423.

85 Heilbronn, H., "On the averages of some arithmetical
    functions of two variables," Mathematika, 5 (1958),
    1-7.

86   Hlawka, E., "Bemerkungen zum grossen Sieb von Linnik,"
     Österreich. Akad. Wiss. Math. -Natur. Kl. S. -B.
     II, 178 (1970), 13-18.

87   Hoheisel, Guido, "Nullstellenanzahl und Mittelwerte der
     Zetafunktion," Sitz. Preuss. Akad. Wiss., 2
     (1930), 1-13.

88   Hoheisel, Guido, "Primzahlprobleme in der Analysis,"
     Sitz. Preuss. Akad. Wiss., 33 (1930), 3-11.

89   Hooley, C., "On the representation of a number as the
     sum of two squares and a prime," Acta Math.,
     97 (1957), 189-210.

90   Hooley, C., "On Artin's conjecture," J. Reine Angew. Math.,
     225 (1967), 209-220.

91   Huxley, M.N., "The large sieve inequality for algebraic
     number fields," Mathematika, 15 (1968), 178-187.

92   Huxley, M.N., "On the differences of primes in
     arithmetical progressions," Acta Arith., 15 (1969),
     367-392.

93   Huxley, M.N., "The large sieve inequality for algebraic
     number fields II: Means of moments of Hecke
     zeta-functions," Proc. London Math. Soc., (3)
     2 (1970), 108-128.

94   Huxley, M.N., "Irregularity in sifted sequences," to
     appear.

95   Iglina, G.S., "On the density of zeros of the zeta-
     function and L-functions near the straight line
     $\sigma = 1$," Izv. Vysš. Učebn. Saved. Matematika,
     55 (1966), 64-73.

96   Ingham, A.E., "Mean-value theorems in the theory of the
     Riemann zeta-function," Proc. London Math. Soc.,
     (2) 27 (1926), 273-300.

97   Ingham, A.E., "On the difference between consecutive primes,"
     Quart. J. Math. Oxford Ser., 8 (1937), 255-266.

98   Ingham, A.E., "On the estimation of $N(\sigma, T)$," Quart. J.
     Math. Oxford Ser., 11 (1940), 291-292.

99   Iseki, Kanesiroo, "A divisor problem concerning prime
     numbers," Japan. J. Math., 21 (1951), 67-92.

100 Joshi, Padmini Tryambak, "The size of $L(1, \chi)$ for real
     non-principal residue characters χ with prime
     modulus," J. Number Theory, 2 (1970), 58-73

101 Jutila, M., "A statistical density theorem for L-functions with applications," Acta Arith., 16 (1969), 207-216.

102 Klimov, N.I., "Almost prime numbers," Uspehi Mat. Nauk, 16 (3) (99) (1961), 181-188. See also: Amer. Math. Soc. Transl., (2) 46 (1965), 48-56.

103 Kober, H., "Eine Mittelwertformel der Riemannschen Zetafunktion," Compositio Math., 3 (1936), 174-189.

104 Landau, Edmund, Handbuch der Lehre von der Verteilung der Primzahlen, B.G. Teubner, Leipzig und Berlin, 1909.

105 Landau, E., "Abschatzungen von Charactersummen, Einheiten und Klassenzahlen," Nachr. Akad. Wiss Göttingen Math. -Phys. Kl. II, (1918), 79-97.

106 Landau, E., "Über die Nullstellen der Dirichletschen Reihen und der Riemannschen ʃ-funktion," Arkiv för Mat. Astronom. och Fysik, 16 (1921), No. 7.

107 Landau, E., "Über die Wurzeln der Zetafunktion," Math. Z., 20 (1924), 98-104.

108 Lavrik, A.F., "On the twin prime hypothesis of the theory of primes by the method of I.M. Vinogradov," Dokl. Akad. Nauk SSSR, 132 (1960), 1013-1015. See also: Soviet Math. Dokl., 1 (1960), 700-702.

109 Lavrik, A.F., "On the distribution of k-twin primes," Dokl. Akad. Nauk SSSR, 132 (1960), 1258-1260. See also: Soviet Math. Dokl., 1 (1960), 764-766.

110 Lavrik, A.F., "The number of k-twin primes lying on an interval of a given length," Dokl. Akad. Nauk SSSR, 136 (1961), 281-283. See also: Soviet Math. Dokl., 2 (1961), 52-55.

111 Lavrik, A.F., "On the theory of the distribution of sets of primes with given differences between them," Dokl. Akad. Nauk SSSR, 138 (1961), 1287-1290. See also: Soviet Math. Dokl., 2 (1961), 827-830.

112 Lavrik, A.F., "Binary problems of additive number theory connected with the method of trigonometric sums of I.M. Vinogradov," Vestnik Leningrad. Univ., 16 (1961), no. 13, 11-27.

113   Lavrik, A.F., "On the theory of distribution of primes based on I.M. Vinogradov's method of trigonometric sums," Trudy Mat. Inst. Stecklov., 64 (1961), 90-125.

114   Lavrik, A.F., "The sum over characters of powers of the modulus of the Dirichlet L-functions in the critical strip," Dokl. Akad. Nauk SSSR, 154 (1964), 34-37. See also: Soviet Math. Dokl., 5 (1964), 28-31.

115   Lavrik, A.F., "An abbreviated approximate functional equation for the functions of Dirichlet," Izv. Akad. Nauk UzSSR Ser. Fiz. -Mat. Nauk, 9 (1965), no. 4, 17-22.

116   Lavrik, A.F., "Functional equations of Dirichlet functions," Dokl. Akad. Nauk SSSR, 17 (1966), 278-280.   See also: Soviet Math. Dokl., 7 (1966), 1471-1473.

117   Lavrik, A.F., "The functional equation for Dirichlet L-functions and the problem of divisors in arithmetic progressions," Izv. Akad. Nauk SSSR Ser. Mat., 30 (1966), 433-448.

118   Lavrik, A.F., "Functional equations of the Dirichlet functions," Izv. Akad. Nauk SSSR Ser. Mat., 31 (1967), 431-442.

119   Lavrik, A.F., "The approximate functional equation for Dirichlet L-functions," Trudy Moscov. Mat. Obšč., 18 (1968), 91-104.

120   Lavrik, A.F., "On L($1,\mathcal{X}$) with real Dirichlet character on sparse sets of values of the modulus of the character," Dokl. Akad. Nauk SSSR, 190 (1970), 1286-1288. See also: Soviet Math. Dokl., 11 (1970), 288-291.

121   Levinson, Norman, "Zeros of the Riemann zeta-function near the 1-line," J. Math. Anal. Appl., 25 (1969), 250-253.

122   Linnik, Yu. V., "The large sieve," Dokl. Akad. Nauk SSSR, 30 (1941), 292-294.

123   Linnik, Yu. V., "On the connection of extended Riemann hypothesis with I.M. Vinogradov's method in the theory of primes," Dokl. Akad. Nauk SSSR, 41 (1943), 145-146.

124   Linnik, Yu. V., "On Dirichlet's L-series and prime number sums," Mat. Sb., 15 (57) (1944), 3-12.

125 Linnik, Yu. V., "On the least prime in an arithmetic progression, I, The basic theorem," Mat. Sb., 15 (57) (1944), 139-178.

126 Linnik, Yu. V., "On the possibility of a unique method in certain problems of "additive" and "distributive" prime number theory," Dokl. Akad. Nauk SSSR, 49 (1945), 3-7.

127 Linnik, Yu. V., "A new proof of the Goldbach-Vinogradov theorem," Mat. Sb., 19 (61) (1946), 3-8.

128 Linnik, Yu. V., "On the density of zeros of L-series," Izv. Akad. Nauk SSSR Ser. Mat., 10 (1946), 35-46.

129 Linnik, Yu. V., "An application of the theory of matrices and of Lobatschevskian geometry to the theory of Dirichlet's real characters," J. Indian Math. Soc., 20 (1956), 37-45.

130 Linnik, Yu. V., "All large numbers are the sums of two squares and a prime (a problem of Hardy and Littlewood), II," Mat. Sb., 53 (95) (1961), 3-38. See also: Amer. Math. Soc. Transl., (2) 37 (1964), 197-240.

131 Linnik, Yu. V., The dispersion method in binary additive problems, American Mathematical Society, Providence, 1963.

132 Linnik, Yu. V., and A.I. Vinogradov, "Estimate of the sum of the number of divisors in a short segment of an arithmetic progression," Uspehi Mat. Nauk, 12 (76) (1957), no. 4, 277-280.

133 Lint, J.H. van, and H. -E. Richert, "Über die summe $\sum \frac{\mu^2(n)}{\varphi(n)}$," Nederl. Akad. Wetensch. Proc. Ser. A, 67 (1964), (= Indag. Math., 26 (1964)), 582-587.

134 Lint, J.H. van, and H.-E. Richert, "On primes in arithmetic progressions," Acta Arith., 11 (1965), 209-216.

135 Littlewood, J.E., "Researches in the theory of the Riemann $\int$-function," Proc. London Math. Soc., (2) 20 (1922), Records, xxii-xxvii.

136 Littlewood, J.E., "On the zeros of the Riemann zeta-function," Proc. Cambridge Philos. Soc., 22 (1924), 295-318.

137 Littlewood, J.E., A Mathematician's Miscellany, Methuen and Co., Ltd., London, 1953.

138 Marcus, Marvin, and Henryk Minc, A survey of matrix
     theory and matrix inequalities, Allyn and Bacon,
     Boston, 1964.

139 Min, S.H., "On the order of $\mathfrak{Z}(\frac{1}{2} + it)$," Trans. Amer.
     Math. Soc., 65 (1949), 448-472.

140 Ming-Chit Liu, "On a result of Davenport and Halberstam,'
     J. Number Theory, 1 (1969), 385-389.

141 Montgomery, H.L., "A note on the large sieve," J. London
     Math. Soc., 43 (1968), 93-98.

142 Montgomery, H.L., "Mean and large values of Dirichlet
     polynomials," Invent. Math., 8 (1969), 334-345.

143 Montgomery, H.L., "Zeros of L-functions," Invent. Math.,
     8 (1969), 346-354.

144 Montgomery, H.L., "Primes in arithmetic progressions,"
     Michigan Math. J., 17 (1970), 33-39.

145 Moroz, B.Z., "Continuability of the scalor product of
     Hecke's series for two quadratic fields," Dokl.
     Akad. Nauk SSSR, 155 (1964), 1265-1267.   See
     also: Soviet Math. Dokl., 5 (1964), 573-575.

146 Moroz, B.Z., "The zeta functions of algebraic number
     fields," Mat. Zametki, 4 (1968), 333-339.

147 Ostrowski, A., "Über das Nichtverschwinden einer Klasse
     von Determinanten und die Lokalisierung der
     charakteristischen Wurzeln von Matrizen,"
     Compositio Math., 9 (1951), 209-226.

148 Paley, R.E.A.C., "On the k-analogues of some theorems in
     the theory of the Riemann zeta-function," Proc.
     London Math. Soc., (2) 32 (1931), 273-311.

149 Parker, W.V., "The characteristic roots of a matrix,"
     Duke Math. J., 3 (1937), 484-487.

150 Perron, O., Theorie der algebraischen Gleichungen, II
     (zweite Auflage), de Gruyter, Berlin, 1933.

151 Pólya, Georg, "Über die Verteilung der quadratischen
     Reste und Nichtreste," Nachr. Akad. Wiss.
     Göttingen Math. -Phys., (1918), 21-29.

152 Pospeev, V.E., "On the 'density' of the zeros of
     Dirichlet L-functions," Izv. Akad. Nauk UzSSR
     Ser. Fiz. -Mat. Nauk, 12 (1968), no. 4, 22-26.

153 Potter, H.S.A., and E.C. Titchmarsh, "The zeros of
     Epstein's zeta-functions," Proc. London Math.
     Soc., (2) 39 (1935), 372-384.

154 Prachar, Karl, Primzahlverteilung, Springer, Berlin, 1957.

155 Rényi, A., "On the representation of an even number as
     the sum of a prime and of an almost prime,"
     Izv. Akad. Nauk SSSR Ser. Mat., 12 (1948), 57-78.
     See also: Amer. Math. Soc. Transl., (2) 19
     (1962), 299-321.

156 Rényi, A., "Un nouveau théorème concernant les fonctions
     indépendantes et ses applications à la théorie
     des nombres," J. Math. Pures Appl., 28 (1949),
     137-149.

157 Rényi, A., "Sur un théorème général de probabilité,"
     Ann. Inst. Fourier, 1 (1950), 43-52.

158 Rényi, A., "On the large sieve of U.V. Linnik,"
     Compositio Math., 8 (1950), 68-75.

159 Rényi, A., "On a general theorem in probability theory
     and its application in the theory of numbers,"
     Zprávy o společnem 3. sjezdu matematikařu
     československých a 7. sjezdu matematikařu
     polskich, Praha, (1950), 167-174.

160 Rényi, A., "On the probabilistic generalization of the
     large sieve of Linnik," Magyar Tud. Akad. Mat.
     Kutató Int. Közl., 3 (1958), 199-206.

161 Rényi, A., "New version of the probabilistic generalization
     of the large sieve," Acta Math. Acad. Sci. Hungar.,
     10 (1959), 217-226.

162 Richert, H.-E., "Zur Abschätzung der Riemannschen
     Zetafunktion in der Nähe der Vertikalen $\sigma = 1$,"
     Math. Ann., 169 (1967), 97-101.

163 Richert, H.-E., "Selberg's sieve with wieghts,"
     Mathematika, 16 (1969), 1-22.

164 Richert, H.-E., "Selberg's sieve with weights," Proceedings
     of the Stony Brook Number Theory Conference, (1969),
     American Mathematical Society, Providence, to
     appear.

165 Rieger, G.J., "Zum Sieb von Linnik," Arch. Math., 11
     (1960), 14-22.

166 Rieger, G.J., "Das grosse Sieb von Linnik für algebraische
     Zahlen," Arch. Math., 12 (1961), 184-187.

167 Rodosskiĭ, K.A., "On the complex zeros of Dirichlet's
        L-functions," Izv. Akad. Nauk SSSR Ser. Mat.,
        12 (1948), 47-56.

168 Rodosskiĭ, K.A., "On the zeros of Dirichlet L-functions,"
        Izv. Akad. Nauk SSSR Ser. Mat., 13 (1949), 315-
        328.

169 Rodosskiĭ, K.A., "On the number of L-functions having
        zeros in a certain rectangle," Ukrain. Mat. Ž.,
        3 (1951), 399-403.

170 Rodosskiĭ, K.A., "On the number of zeros of all
        L-functions with characters of given modulus,"
        Dokl. Akad. Nauk SSSR, 84 (1952), 669-671.

171 Rodosskiĭ, K.A., "On the least prime number in an
        arithmetic progression and the zeros of
        L-functions," Dokl. Akad. Nauk SSSR, 88 (1953),
        753-756.

172 Rodosskiĭ, K.A., "On the least prime in an arithmetic
        progression," Mat. Sb., 34 (76) (1954), 331-356.

173 Rodosskiĭ, K.A., "On non-residues and zeros of L-functions,"
        Izv. Akad. Nauk SSSR Ser. Mat., 20 (1956), 303-306.

174 Rodriguez, G., "Sul problema dei divisori di Titchmarsh,"
        Boll. Un. Mat. Ital., (3) 20 (1965), 358-366.

175 Rosser, J. Barkley, and Lowell Shoenfeld, "Approximate
        formulas for some functions of prime numbers,"
        Illinois J. Math., 6 (1962), 64-94.

176 Roth, K.F., "Remark concerning integer sequences," Acta
        Arith., 9 (1964), 257-260.

177 Roth, K.F., "On the large sieves of Linnik and Rényi,"
        Mathematika, 12 (1965), 1-9.

178 Roth, K.F., "Irregularities of sequences relative to
        arithmetic progressions," Math. Ann., 169 (1967),
        1-25.

179 Roth, K.F., "The large sieve," Inaugural Lecture, 23
        January, 1968, Imperial College of Science and
        Technology, London, 1968, 9 pp.

180 Roth, K.F., "Irregularities of sequences relative to
        arithmetic progressions, III," J. Number Theory,
        2 (1970), 125-142.

181 Salié, Hans, "Über den kleinsten positiven quadratischen
     Nichtrest nach einer Primzahl," Math. Nachr.,
     3 (1949), 7-8.

182 Samandarov, A.G., "On the large sieve in algebraic number
     fields," Mat. Zametki, 2, 6 (1967), 673-680.

183 Schaal, Werner, "On the large sieve method in algebraic
     number fields," J. Number Theory, 2 (1970),
     249-270.

184 Schinzel, A., "Remarks on the paper 'Sur certaines
     hypothèses concernant les nombres premiers,'"
     Acta Arith., 11 (1961), 1-8.

185 Schinzel, A., and W. Sierpiński, "Sur certaines hypo-
     thèses concernant les nombres premier," Acta
     Arith., 4 (1958), 185-208.

186 Schur, J., "Einige Bemerkungen zu der vorstehenden
     Arbeit des Herrn. G. Pólya: Über die Verteilung
     der quadratischen Reste und Nichtreste," Nachr.
     Akad. Wiss. Göttingen Math. Phys. Kl., (1918),
     30-36.

187 Selberg, A., "On the zeros of Riemann's zeta-function,"
     Skr. Norske Vid. -Akad. Oslo I, (1942), no. 10,
     59 pp.

188 Selberg, A., "On the normal density of primes in small
     intervals, and the difference between consec-
     utive primes, Arch. Math. Naturvid., 47 (1943),
     no. 6, 87-105.

189 Selberg, A., "On the remainder formula for N(T)," Skr.
     Norske Vid. -Akad. Oslo I, (1944), no. 1, 27 pp.

190 Selberg, A., "Contributions to the theory of Dirichlet's
     L-functions," Skr. Norske Vid. -Akad. Oslo I,
     (1946), no. 3, 62 pp.

191 Selberg, A., "Contributions to the theory of the Riemann
     zeta-function," Arch. Math. Naturvid., 48 (1946),
     89-155.

192 Selberg, A., "The general sieve-method and its place in
     prime number theory," Proceedings of the
     International Mathematical Congress, 1950, vol. 1,
     286-292.

193 Selberg, A., "On elementary methods in prime number theory
     and their applications," Den 11te Skandinaviske
     Matematikerkongress, Trondheim, (1949), (Johan
     Grundt Tanums Forlag, Oslo, 1952), 13-22.

194 Selberg, A., "Sieve methods," Proceedings of the Stony
      Brook Number Theory Conference, 1969, American
      Mathematical Society, Providence, to appear.

195 Shanks, Daniel, "On maximal gaps between successive
      primes," Math. Comp., 18 (1964), 646-651.

196 Siegel, C.L., "Über Gitterpunkte in convexen Körpern
      und ein damit zusammenhängendes Extremalproblem,"
      Acta Math., 65 (1935), 307-323.

197 Tatuzawa, Tikao, "On the zeros of Dirichlet's L-functions,"
      Proc. Japan Acad., 26 (1950), no. 9, 1-13.

198 Titchmarsh, E.C., "The mean-value of the zeta-function on
      the critical line," Proc. London Math. Soc., (2)
      27 (1928), 137-150.

199 Titchmarsh, E.C., "On the zeros of the Riemann zeta-
      function," Proc. London Math. Soc., (2) 30 (1929),
      319-321.

200 Titchmarsh, E.C., "A divisor problem," Rend. Circ. Mat.
      Palermo, 54 (1930), 414-429.

201 Titchmarsh, E.C., "on van der Corput's method and the zeta
      function of Riemann (V)," Quart. J. Math. Oxford
      Ser., 5 (1934), 195-210.

202 Titchmarsh, E.C., "The mean value of $|\mathfrak{Z}(\frac{1}{2} + it)|^4$," Quart.
      J. Math. Oxford Ser., 8 (1937), 107-112.

203 Titchmarsh, E.C., The theory of functions, Second Ed.,
      Oxford University Press, Oxford, 1939.

204 Titchmarsh, E.C., "On the order of $\mathfrak{Z}(\frac{1}{2} + it)$," Quart. J.
      Math. Oxford Ser., 13 (1942), 11-17.

205 Titchmarsh, E.C., The theory of the Riemann zeta-function,
      Clarendon Press, Oxford, (1951).

206 Turán, Paul, "Über die Primzahlen der arithmetischen
      Progression," Acta Sci. Math. (Szeged), 8 (1937),
      226-235.

207 Turán, P., "Über die Verteilung der Primzahlen (I),"
      Acta. Sci. Math. (Szeged), 10 (1941), 84-104.

208 Turán, P., "Über die Wurzeln der Dirichletschen
      L-funktionen," Acta Sci. Math. (Szeged), 10
      (3-4) (1943), 188-201.

209   Turán, P., "On Carlson's theorem in the theory of the
       zeta-function of Riemann," Acta Math. Acad. Sci.
       Hungar., 2 (1951), 39-73.

210   Turán, P., "On the roots of the Riemann zeta-function,"
       Magyar Tud. Akad. Mat. Fiz. Oszt. Közl, 4 (1954),
       357-368.

211   Turán, P., "On Lindelöf's conjecture," Acta Math. Acad. Sci.
       Hungar., 5 (1954), 145-163.

212   Turán, P., "On the zeros of the zeta function of Riemann,"
       J. Indian Math. Soc., 20 (1956), 17-36.

213   Turán, P., "On the so-called density-hypothesis in the
       theory of the zeta-function of Riemann," Acta
       Arith., 4 (1958), 31-56.

214   Turán, P., "On a density theorem of Yu. V. Linnik,"
       Publ. Math. Inst. Hungar. Acad. Sci., 6 (1961),
       165-179.

215   Turán, P., "On the twin-prime problem, I," Publ. Math.
       Inst. Hungar, Acad. Sci., 9 (3-A) (1964),
       247-261.

216   Turán, P., "Some function-theoretic sieve methods in the
       theory of numbers," Dokl. Akad. Nauk SSSR, 171
       (1966), 1661-1664.

217   Turán, P., "On the twin-prime problem, II," Acta Arith.,
       13 (1967), 61-89.

218   Turán, P., "On the twin prime problem, III," Acta Arith.,
       14 (1968), 399-407.

219   Vinogradov, A.I., "On the continuability into the left
       half-plane of the scalor product of Hecke
       L-series with Grössencharaktere," Izv. Akad. Nauk
       SSSR Ser. Mat., 29 (1965), 485-492.

220   Vinogradov, A.I., "On the density hypothesis for
       Dirichlet L-functions," Izv. Akad. Nauk SSSR Ser.
       Mat., 29 (1965), 903-934. See Also: "Correction
       to the paper of A.I. Vinogradov, 'On the density
       hypothesis for Dirichlet L-functions,'" Izv. Akad.
       Nauk SSSR Ser. Mat., 30 (1966), 719-720.

221   Vinogradov, I.M., "Sur la distribution des résidus et non
       résidus de puissances," Permski J. Phys. Isp.
       Ob.-wa, 1 (1918), 18-28, 94-98.

222 Vinogradov, I.M., "Representation of an odd number as a
       sum of three primes," Dokl. Akad. Nauk SSSR,
       15 (1937), 169-172.

223 Vinogradov, I.M., "Some theorems concerning the theory
       of primes," Mat. Sb., 2 (44) 2 (1937), 179-195.

224 Vinogradov, I.M., The method of trigonometrical sums
       in the theory of numbers, Interscience
       Publishers, New York.

225 Walfisz, A., "Zur Abschätzung von $\mathfrak{f}(\tfrac{1}{2} + it)$," Nachr.
       Akad. Wiss. Göttingen Math. -Phys. Kl., (1924),
       155-158.

226 Walfisz, A., Weylsche Exponentialsummen in der neueren
       Zahlentheorie, VEB Deutcher Verlag der Wiss.,
       Berlin, 1963.

227 Wilson, R.J., "The large sieve in algebraic number fields,"
       Mathematika, 16 (1969), 189-204.